THE NATIONAL ACADEMIES

Advisers to the Nation on Science, Engineering, and Medicine

The **National Academy of Sciences** is a private, nonprofit, self-perpetuating society of distinguished scholars engaged in scientific and engineering research, dedicated to the furtherance of science and technology and to their use for the general welfare. Upon the authority of the charter granted to it by the Congress in 1863, the Academy has a mandate that requires it to advise the federal government on scientific and technical matters. Dr. Bruce M. Alberts is president of the National Academy of Sciences.

The **National Academy of Engineering** was established in 1964, under the charter of the National Academy of Sciences, as a parallel organization of outstanding engineers. It is autonomous in its administration and in the selection of its members, sharing with the National Academy of Sciences the responsibility for advising the federal government. The National Academy of Engineering also sponsors engineering programs aimed at meeting national needs, encourages education and research, and recognizes the superior achievement of engineers. Dr. Wm. A. Wulf is president of the National Academy of Engineering.

The **Institute of Medicine** was established in 1970 by the National Academy of Sciences to secure the services of eminent members of appropriate professions in the examination of policy matters pertaining to the health of the public. The Institute acts under the responsibility given to the National Academy of Sciences by its congressional charter to be an adviser to the federal government and, upon its own initiative, to identify issues of medical care, research, and education. Dr. Harvey V. Fineberg is president of the Institute of Medicine.

The **National Research Council** was organized by the National Academy of Sciences in 1916 to associate the broad community of science and technology with the Academy's purposes of furthering knowledge and advising the federal government. Functioning in accordance with general policies determined by the Academy, the Council has become the principal operating agency of both the National Academy of Sciences and the National Academy of Engineering in providing services to the government, the public, and the scientific and engineering communities. The Council is administered jointly by both Academies and the Institute of Medicine. Dr. Bruce M. Alberts and Dr. Wm. A. Wulf are chair and vice-chair, respectively, of the National Research Council.

www.national-academies.org

ANALYTICAL METHODS AND APPROACHES FOR WATER RESOURCES PROJECT PLANNING

Panel on Methods and Techniques of Project Analysis

Committee to Assess the U.S. Army Corps of Engineers Methods of Analysis and Peer Review for Water Resources Project Planning

Water Science and Technology Board

Ocean Studies Board

Division on Earth and Life Studies

NATIONAL ACADEMY OF SCIENCES
THE NATIONAL ACADEMIES

THE NATIONAL ACADEMIES PRESS
Washington, D.C.
www.nap.edu

THE NATIONAL ACADEMIES PRESS 500 Fifth Street, N.W. Washington, DC 20001

NOTICE: The project that is the subject of this report was approved by the governing board of the National Research Council, whose members are drawn from the councils of the National Academy of Sciences, the National Academy of Engineering, and the Institute of Medicine. The members of the committee responsible for the report were chosen for their special competences and with regard for appropriate balance.

Support for this project was provided by the U.S. Army Corps of Engineers under contract no. DACW72-01-C-0001. Any opinions, findings, conclusions, or recommendations expressed in this publication are those of the author(s) and do not necessarily reflect the view of the organizations or agencies that provided support for the project.

International Standard Book Number 0-309-09182-9 (Book)
International Standard Book Number 0-309-53128-4 (PDF)

Analytical Methods and Approaches for Water Resources Project Planning is available from the National Academies Press, 500 Fifth Street, N.W. Lockbox 285, Washington, DC 20055; (800) 624-6242 or (202) 334-3313 (in the Washington metropolitan area); Internet, http://www.nap.edu.

Cover design by Van Nguyen, the National Academies Press. Photo courtesy of the Missouri Historical Society. Photo as cited in Shallat, T. *Structures in the Stream.*

Printed in the United States of America.

COMMITTEE TO ASSESS THE U.S. ARMY CORPS OF ENGINEERS METHODS OF ANALYSIS AND PEER REVIEW FOR WATER RESOURCES PROJECT PLANNING

PANEL ON METHODS AND TECHNIQUES OF PROJECT ANALYSIS*

GREGORY B. BAECHER, *Chair*, University of Maryland, College Park
JOHN B. BRADEN, University of Illinois, Urbana-Champaign
DAVID L. GALAT, University of Missouri, Columbia
GERALD E. GALLOWAY, Titan Corporation, Fairfax, Virginia
ROBERT G. HEALY, Duke University, Durham, North Carolina
EDWIN E. HERRICKS, University of Illinois, Urbana-Champaign
CATHERINE L. KLING, Iowa State University, Ames
LINDA A. MALONE, College of William and Mary, Williamsburg, Virginia
RAM MOHAN, Blasland, Bouck & Lee, Inc., Annapolis, Maryland
MAX J. PFEFFER, Cornell University, Ithaca, New York
DOUG PLASENCIA, AMEC, Phoenix, Arizona
DENISE J. REED, University of New Orleans, Louisiana
JAN A. VELTROP, Consultant, Skokie, Illinois

National Research Council Staff

JEFFREY W. JACOBS, Study Director
ELLEN A. DE GUZMAN, Research Associate

* The Panel on Methods and Techniques of Project Analysis was one of four panels, operating under the auspices of a coordinating committee that was convened by the National Academies' Water Science and Technology Board (lead) and Ocean Studies Board to carry out studies mandated in the Water Resources Development Act of 2000. The members of the two boards and the panels are listed in Appendix F.

Foreword

In the early 1800s the U.S. Congress first asked the U.S. Army Corps of Engineers (which was created in 1775) to improve navigation on our waterways. From that beginning, the Corps began a program of public works that has reshaped virtually all of the nation's river basins and coastal areas. Today we share in the benefits of those works: a reliable water transportation network, harbors that help link our economy to global markets, previously flood-prone land that is productive for urban and agricultural uses, hydroelectric power, and widely used recreational facilities.

Now, at the beginning of the twenty-first century, the Corps' program is under intense scrutiny. Traditional constituencies press the Corps to complete projects that have been planned for many years and campaign for new projects to serve traditional flood control and navigation purposes. At the same time, environmental and taxpayer groups express concerns about these projects in Congress and in the courts. Some of these groups have exposed technical errors in analyses that have been used to justify projects. For these critics, the Corps' water project development program must be reformed and the budget reduced or redirected.

Some of these same groups are pressing the administration, Congress, and the agency itself toward a new Corps mission, broadly described as environmental restoration. However, the concept of restoration awaits more precise definition, and the science of ecosystem restoration is in its infancy. Nevertheless, it is clear that restoration is a call for water resources management that accommodates and benefits from, rather than controls, annual and multiyear variability in the patterns and timing of river flows and the extremes of flood and drought.

Meanwhile, the Corps is affected by a general trend in all federal agencies toward smaller budgets and staffs. As demands for reform mount, the Corps' current staffing and organization may have to be reconfigured to provide improved and more credible planning reports.

As a result of this national debate over the Corps' programs and the quality of its planning studies, the U.S. Congress in Section 216 of the

2000 Water Resources Development Act, requested that the National Academies conduct a study of procedures for reviewing the Corps' planning studies. In addition, Congress requested a review of the "methods of analysis" used in Corps water resources planning.

In response to this request, the Water Science and Technology Board of the National Academies' National Research Council (NRC), in collaboration with the NRC's Ocean Studies Board, appointed four study panels to assess (1) peer review, (2) planning methods, (3) river basin and coastal systems planning, and (4) resource stewardship and adaptive management, along with a coordinating committee to follow these panels' progress and to write a synthesis report.

Our study panels and coordinating committee held several meetings over the course of the study period beginning in 2001. We spoke with dozens of Corps of Engineers personnel, visited several Corps projects, and heard from different groups with interests in Corps projects. We came away with an appreciation for the dedication of Corps personnel and the complications and challenges they face in trying to be responsive to local project sponsors and the nation's taxpayers.

This is not the first study of the Corps by the National Academies. However, past studies were often focused on specific projects or on particular planning aspects. The reports in this series address the agency's programs in a wider context. Because we appreciate the importance of the U.S. Congress and the sitting administration in directing Corps programs, many of our recommendations are directed to them.

The Corps has a long history of serving the nation and is one of our oldest and most recognized federal agencies, but it is today at an important crossroads. The nation, through the administration and Congress, must help the agency chart its way for the next century.

Leonard Shabman
Chair, Coordinating Committee

Preface

The U.S. Army Corps of Engineers has been the leading federal agency for inland navigation improvement, flood protection, coastal works, and other aspects of U.S. water resources management essentially since the founding of the republic. The Corps has gone through several phases in its extensive history. During the early twentieth century, there was a focus on enhancing navigability of the nation's rivers. Following the devastating 1927 Mississippi River floods, better management of flood risks was a high priority. During the middle of the twentieth century, concepts of multi-purpose projects and river basin planning were viewed as central means for promoting economic development, among other goals. The close of the twentieth century saw the Corps becoming involved in ecosystem restoration projects, such as the Everglades restoration project. Specific water-related events and disasters, such as the 1993 Mississippi and Missouri River floods, resulted in policy challenges and reconsiderations of national water management strategies. Through its history, the Corps' mission has been to carry out the federal interest as manifest in congressional direction, and in so doing, to evaluate potential water resources projects against the backdrop of national goals.

Reviews of Corps of Engineers project analysis and evaluation must consider this long history of project development, the complex relations of the Corps to other federal agencies with water resources management duties, historical controversies that have attended the Corps' execution of its missions, and the breadth of water resources activities and projects overseen by the Corps. Our study panel was not the first group to review Corps of Engineers and federal water resources roles and planning procedures. Previous statements on federal water planning were issued by, for example, the National Resources Committee (1938), the Hoover Commission (1949), the Cooke Commission (1950), the Kerr Committee of the Senate (1959), and the National Water Commission (1968 and 1973), and by previous committees of National Research Council (e.g., NRC, 1999). Although specific planning techniques and larger national priorities shifted through the course of these studies, several overarching themes stand out: the importance of the executive and congressional branches in providing clear policy direction, the importance of clarifying the respective roles of the various federal and state agencies involved in water resources planning, the importance of financial and economic decision criteria in planning, and the importance of sound technical bases of economic and engineering evaluation.

These issues remain important in today's national water resources planning context. Our panel's task was to review Corps planning techniques as embodied in the federal *Principles and Guidelines* and in the Corps own planning guidance. In our discussions, it soon became evident that comprehensive evaluation of these topics had to be framed by the setting in which planning decisions are made. These contextual issues include national water resource policy guidance for federal agency decision making, multi-agency and federal-state cooperation, accounting for multi-attribute outcomes in project evaluations, roles of local stakeholders in federally-sponsored projects, and accommodation of risk and uncertainty in project planning. Water resources project planning by the Corps of Engineers does not occur in a vacuum. The Corps depends on clear guidance from the Congress, and the Corps must work in partnership with local stakeholders to help understand and uphold the federal interest in water project planning and investments.

Our panel is grateful to the many representatives of the U.S. Army Corps of Engineers, both in Washington and in the Districts, who offered information and insights during the course of this study. James Johnson and Harry Kitch from Corps Headquarters provided important historical and analytical insights into Corps planning procedures. Former Assistant Secretary of the Army for Civil Works, Dominic Izzo, shared his views on Corps missions, structure, and strategies with our panel. David Moser and Eugene Stakhiv of the Institute for Water Resources outlined for us the new directions in economic analysis being developed for future Corps projects. Other Corps of Engineers military and civilian staff members who spoke with us include Colonel Robert Ball, Buddy Arnold, Robert Lindner, Ed McNally, Marsha Mose, and Susan Smith. Several other Corps staff members, including several from the Corps St. Paul District who hosted a field trip along the Mississippi River in St. Paul, graciously shared their time and insights. We also thank Steve Fitzgerald and John Williams, who spoke with us at our first meeting and shared their perspectives as a local Corps project sponsor and as an independent analyst of Corps methods, respectively. We are similarly grateful to many members of the public and to representatives of professional and public policy organizations who generously spent time meeting with the committee or who provided us with documents for our consideration. The collective input of these individuals and organizations was critical to our group's education.

The panel, and particularly the panel chairman, are grateful to the NRC staff for its dedication and diligence in organizing the committee's activities and in helping bring this report to fruition. Our work would not have been possible without the excellent staff of the Water Science and Technology Board. Stephen Parker, Director of the Board, provided advice and direction for the overarching "Section 216" studies. Jeffrey Jacobs, Senior Staff Officer of the Board, spent many days and weekends crafting the sometimes less-than-splendid prose of the members and committee chairman into a concise and articulate form. They were ably assisted by Ellen de Guzman, Research Associate at the Water Science and Technology Board, who deftly handled administrative details for meetings and supported all phases of report preparation, including editorial and graphics work. Leonard Shabman, Chairman of the Coordinating Committee for the Section 216 Studies, also merits special mention, as he provided an abundance of constructive and tactful advice and suggestions and served as a continuing source of insight for the panel's deliberations.

The report was reviewed in draft form by individuals chosen for diversity of perspective and technical expertise in accordance with the procedures approved by the NRC's Report Review Committee. The purpose of this independent review is to provide candid and critical comments that will assist the institution in making its published report as sound as possible and to ensure that the report meets institutional standards for objectivity, evidence, and responsiveness to the study charge. The review comments and draft manuscript remain confidential to protect the integrity of the deliberative process. We thank the following for their review of this report: Charles Howe, University of Colorado (ret.); David Kennedy, California Department of Water Resources (ret.); Jon Kusler, Association of State Floodplain Managers; Walter Lynn, Cornell University (ret.); David Moreau, University of North Carolina; Herbert Ward, Rice University; and Douglas Woolley, Radford University. Although these reviewers provided many constructive comments and suggestions, they were not asked to endorse the conclusions or the recommendations, nor did they see the final draft of the report before its release. The review of this report was overseen by Leo Eisel of Brown and Caldwell and by Richard Conway (retired) of Union Carbide Corporation. Appointed by the National Research Council, they were responsible for ensuring that an independent examination of the report was carefully carried out in accordance with NRC institutional procedures and that all review comments were carefully considered. Responsibility for the final content of this report rests entirely with the authoring committee and the NRC.

The Corps of Engineers has a long history of leadership in the development and application of analytical methods for water project planning. As federal roles in water management, budgetary priorities, and social preferences all change, the Corps' roles and methods for planning and managing water resources will have to similarly adjust. The Corps of Engineers is responsible for controlling and managing a substantial portion of the nation's hydrologic and related resources. The nation needs a competent and modern Corps of Engineers if these resources are to be managed by sound planning methods and to meet current and future national water management priorities. We offer our report in the spirit of helping the Corps—with support from our elected leaders—to meet these important challenges.

Gregory B. Baecher
Chairman

Contents

Executive Summary

The U.S. Army Corps of Engineers has long constructed civil works projects to help manage the nation's inland and coastal water systems. The Corps' earliest water-related missions were to manage flood risks and enhance navigation. The U.S. Congress added several missions, including beach and hurricane protection, hydropower production, water supply, as well as ecosystem restoration (the latter responsibility was added in the 1990s), to the Corps work program over the years. In promoting these objectives, thc agency constructed dams, levees, and other civil works to help reduce high river flows and flood damages, and to support inland and harbor navigation. Although the Corps has never been without its critics, through the 1960s the agency could generally rely upon widespread congressional and public support for its civil works projects. An area in which the Corps was not traditionally criticized was the quality of the agency's planning and evaluation methods and the technical soundness of its planning studies.

But the planning context has changed greatly since the 1960s, with criticisms of the Corps' basic objectives and of its planning methods becoming increasingly sharp. The agency has made concerted efforts to adapt to these criticisms, and the typical Corps of Engineers planning study today incorporates vast amounts of economics and engineering data, exhaustive environmental analyses, and is several hundreds pages long. These efforts toward increasing sophistication and thoroughness, however, have not silenced the Corps' critics. In fact, the Corps' efforts to improve and refine its planning and analytical techniques seem to have only invited additional scrutiny and criticism. To compound matters, the criticisms come at a time when the national interest in water development is in a state of flux. Moreover, engineering capabilities in the private sector have greatly increased, and the Corps no longer enjoys a once near-monopoly in civil engineering expertise.

This report was part of a larger study that was conducted in response to a request from the U.S. Congress in the Water Resources Development Act of 2000 for the National Academy of Sciences to review the Corps' peer review methods and analytical approaches (this report's Foreword and Chapter 1 discuss these studies). This panel reviewed the Corps' analytical procedures and planning methods, largely in the context of the federal *Economic and Environmental Principles and Guidelines for Water and Related Land Resources Implementation Studies*, also known as the *Principles and Guidelines* or "*P and G"(P&G)*, as well as the Corps' *Planning Guidance Notebook* (*PGN*). This panel's statement of task was:

> This panel will review the formulation and evaluation techniques and methods, including the use of models for project-specific applications (i.e., benefits, costs, optimization, etc.) with a view toward developing and refining planning methods to better serve the nation. As appropriate, the panel should also consider the methods, techniques, and practices employed by other federal agencies, states, and the private sector in the development of the projects.
>
> The panel will also consider the need for system-wide considerations in project-specific planning and other state-of-the-art methods. The need to modernize and/or better implement the Corps methods and other techniques will be evaluated in the context of the current federal Principles and Guidelines (P&G). In addition, the panel will examine the interpretation of the P&G as reflected in Corps guidance such as ER 1105-2-100. This panel will also perform an ex post analysis of a sample of Corps projects (including major project purposes) based on 1) the methods used by the Corps and 2) state-of-the-art methods. As appropriate, the panel will make recommendations for improving Corps methods and techniques and may make recommendations regarding areas of the P&G in need of modernization.

This report provides recommendations for improving the processes and methods outlined in the federal *P&G* and the Corps *PGN*. An observation noted during the course of this study is that substantial improvements to Corps planning methods will only be realized if they are linked to changes in broader, federal-level organizational and policy structures and processes that frame and guide those methods. Before presenting findings and recommendations regarding Corps planning and evaluation methods, it is thus appropriate to review federal-level governance issues relevant to these methods.

FEDERAL LEGISLATION, GUIDANCE, AND ORGANIZATIONS

Policies and Legislation

The contemporary setting of U.S. water management features numerous well-informed and active stakeholders with water management objectives that are often mutually exclusive. In this setting, clear direction from the Administration and the Congress—which have ultimate responsibility for federal water project management decisions—regarding the nation's *de facto* body of water policy, is essential. This clarification is especially important given that the Corps' future work program is likely to center upon more efficient management of existing infrastructure—which will entail more trade-off decisions and less new project construction.

This is not to say that the Corps lacks federal-level guidance. On the contrary, the Corps is governed by 219 public laws (as of December 2000) that date back to the late nineteenth century. Executive orders, congressional committee language, and Administration guidance also guide the Corps. But internal inconsistencies within this body of guidance often pose problems for the Corps, as this body was not designed according to a master plan, but rather accreted over time with little regard given to whether new laws and directives were fully consistent with existing ones. For example, the Corps must abide by environmental legislation such as the National Environmental Policy Act, the Clean Water Act, and the Endangered Species Act. At the same time, however, the Corps is also mandated to provide services such as navigation channels. This results in conflicts when, for example, an endangered species' habitat requirements (e.g., a flow regime that includes high flows, low flows, and occasional overbank floods) clash with an authorization(s) for a navigation channel of a given, reliable depth. Inadequate guidance on how to resolve these types of conflicts places the Corps in the position of setting policy, often resulting in confused policy direction, inefficient operations, and criticism of the agency from all corners.

1. To provide clearer direction to the Corps, the Administration and the Congress, in cooperation with the states, should reconcile inconsistencies within the existing, de facto, body of national water policy.

Coordinating National Water Policy

Resolving inconsistencies within the body of national water policy, and resolving inconsistencies among principles and methods employed by various federal water resources agencies, will require sustained attention, with controversial, high profile cases likely to require careful individual consideration. Given the large number of federal agencies (more than a dozen) with responsibilities for managing water and related resources, meaningful progress on this front will require a body specifically devoted to this task.

2. A body should be specifically charged to coordinate water resources policies and activities among the Administration, the Congress, the States, and federal agencies with water resources management responsibilities.

Revising the *Principles and Guidelines*

The Corps of Engineers' key planning guidance documents are the federal *Principles and Guidelines* (1983) and the Corps' *Planning Guidance Notebook* (2000), which provides guidance consistent with the *P&G*. The *P&G* document was approved in 1983 and is derived in part from the 1972 federal *Principles and Standards*. The *Principles and Guidelines* document, and its *Principles and Standards* predecessor, represented many relevant and useful economic and planning concepts. Nonetheless, the *P&G* is based in part upon water management and development paradigms of the 1970s (in some cases earlier). The *P&G* has not been revised for over twenty years, a period of substantial changes in planning approaches, improvements in analytical methods, and shifts in social preferences and scientific paradigms. The *P&G* document defines the current federal water and related land resources objective: "to contribute to national economic development, consistent with protecting the nation's environment, pursuant to national environmental statutes, applicable executive orders, and other Federal planning requirements" (WRC, 1983). This "National Economic Development" objective is clear, and it likely reflects the contemporary views of many congressional representatives and many of their constituents. Nonetheless, its viability has diminished in today's context of more vigorous stakeholder participation and a broader set of demands from the nation's river and coastal systems.

The *P&G* prescribes a set of detailed methods designed to quantify costs and benefits associated with a proposed water project. The *P&G* relies heavily upon predictive models and monetization techniques as the basis for water resources investment decisions. The theories and methods within the *P&G*, however, do not adequately reflect contemporary planning approaches or settings. The limits of models to provide precise predictions are today better understood and appreciated. Similarly, the limits of monetization techniques—which have realized some notable recent analytical advances—to accurately quantify all the important variables related to a water management decision are also better understood. Stakeholder groups, especially project co-sponsors, today demand a strong voice in important water management decisions. Adaptive management principles are being implemented in some places to help managers adjust to unknown and unforeseen anthropogenic impacts on ecosystem processes. The Corps recognizes these issues, but the *P&G* does not incorporate these types of considerations into the planning calculus.

The Corps naturally looks to credible sources to guide its planning decisions. This is especially the case given that the agency's planning methods and decisions are carefully scrutinized, and that internally inconsistent federal legislation and other policy directives generally do not provide clear project management objectives. Corps planners use the *P&G* as an analytical touchstone to navigate through this context. But a quest for greater credibility through the development and application of increasingly detailed analytical methods has resulted in those methods becoming a product in themselves, rather than serving as a means to reach sound water management decisions.

3. The Principles and Guidelines should be revised to better reflect contemporary management paradigms, analytical methods, legislative directives, and social, economic, and political realities. The new planning guidance should apply to water resources implementation studies and similar evaluations carried out by all federal agencies. A revised version of the P&G document should be periodically and formally reviewed and updated.

No significant action has yet taken place within the Administration in response to this recommendation that has been voiced multiple times by previous groups.

4. Therefore, even if the Administration should choose not to revise the P&G, the Corps should draft a revision to its Planning Guidance Notebook that is consistent with this report's recommendations and propose this revision to the Administration.

The new planning guidance should recognize uncertainties in planning and the limits of predictive models. It should reflect the reality that planning decisions are not purely analytical, that significant uncertainties often exist, and that they are influenced by factors such as stakeholder preferences. The reconciliation of some existing discrepancies among legislation and other guidance would provide clearer guidance for the Corps. This would allow for new planning guidance to be less prescriptive than the *P&G*, which would better accommodate stakeholder input and would encourage the creativity of Corps analysts. The new planning guidance should be periodically reviewed and updated.

A revised *P&G* document will not be a panacea for the Corps, as a revised document will eliminate neither differences between competing interests and values nor eliminate the analytical complexities of planning and managing large water projects. But because the Corps is obliged to follow the *P&G*, because the agency takes the document seriously, and because the *P&G* does not adequately reflect contemporary realities and best practices, its revision should prove useful to the Corps.

CORPS OF ENGINEERS PLANNING PROCESSES

Benefit-Cost Analysis

Pursuant to federal legislation and administration directives, a proposed water project is viable only if its projected benefits exceed projected costs as documented in a Corps planning study. Although this rule imposes discipline on the planning process, it often pressures analysts to make questionable assumptions or to configure a study such that it produces a given benefit-cost ratio. Benefit-cost analysis is an important component of sound decision making. But benefit-cost analysis may not adequately consider uncertainties and relevant public policy considerations such as stakeholder opinions, non-market values, and equity.

5. Benefit-cost analysis should not be used as the lone decision criterion in judging whether a proposed planning or management alternative in a Corps planning study should be approved. Benefit-cost analysis is a useful decision making guide, and it is important that

Corps planning studies retain the discipline imposed by comparing project benefits and costs. But it should be used only as one criterion in reaching final judgment on a proper alternative; criteria such as stakeholder opinions, political preferences, equity, and non-market values such as biodiversity are important factors in water resources investments and policy decisions that are not captured in benefit-cost analysis. Rather than being the product of lengthy analysis (as currently embodied within the *P&G*), benefit-cost analyses in future Corps studies should aim to be more concisely presented, with clear explanations of assumptions and models employed.

Review of Projects and Planning Studies

Ex Post Evaluation

The Corps of Engineers does not systematically review outcomes of its water projects after construction or review how closely planning study projections matched actual results. A lack of retrospective, or "ex post," reviews represents missed opportunities to better understand how demands upon water projects have changed over time, strengths and weaknesses of planning methods, and how project operations have (or have not) changed to meet changing conditions. The Corps has authorities that allow for post-construction evaluations to be conducted and for operations to be adjusted accordingly, but Congress generally has not provided resources to the Corps for comprehensive post-construction studies. Evaluations of water project outcomes, however, are essential to sound water resources management (the report from the 216 study Coordinating Committee provides more detailed advice on the role of study authorities in improving water planning decisions).

6. Periodic reviews of completed projects should be a routine part of Corps water project planning and management. Congress should provide resources to conduct these "ex post" evaluations.

Strengthened Reconnaissance Studies

The Corps of Engineers conducts its planning studies in two phases, a reconnaissance phase and a feasibility phase. The Corps' internal guidelines limit those reconnaissance studies to one year and to $100,000. A reconnaissance study is conducted to determine if a national interest exists in addressing a given water resources problem or opportunity. It also provides an opportunity to consider a broad range of planning alternatives. The given time and resource constraints may be reasonable for smaller, less expensive studies. Larger, more sophisticated studies, however, may require additional time and resources. Inadequate study reconnaissance in important projects may preclude the consideration of viable alternatives, which may create blind spots and contribute to conflicts during the feasibility study.

7. Resources and time allocated for Corps reconnaissance studies should be commensurate with the scale and complexity of the water resources issue at hand.

Stakeholder Participation

The Corps has been a leader in developing and implementing stakeholder participation procedures. Since the late 1980s, however, the Corps has been less systematic in incorporating stakeholder participation into planning and management decisions. Although stakeholder participation is required in all project planning, its design and conduct takes place at the Corps district (local) level, with only general standards grounded in legal requirements like those associated with the National Environmental Policy Act. This has resulted in inconsistency in stakeholder participation practices across projects and Corps offices.

8. The Corps should conduct a comprehensive review of district-level experiences with stakeholder participation procedures and activities. The Corps should also develop training and reference materials on stakeholder participation standards.

Summary Document in Planning Studies

Corps of Engineers planning studies, and attending appendixes and other documents, are often hundreds of pages in length. This quantity of data often makes it difficult to identify and comprehend all important assumptions, alternatives, models employed, data sets, and other factors.

9. A summary document that identifies key environmental and social issues, primary assumptions, alternatives considered and evaluated, objectives sought, benefits and costs (monetized and non-

monetized), trade-offs and stakeholder perspectives and differences, presented with a consistent format across studies, should be a standard in Corps planning studies.

Engineering Methods

Three aspects of the Corps' engineering analysis and methods bear close attention in the years ahead. The importance of these relates to changing paradigms of U.S. water resources management and to changing needs of Corps projects and activities. These are (1) systems engineering aspects of water resource planning, (2) impacts of risk and uncertainty on planning, and (3) integrating engineering methods of analysis with ecosystem restoration planning. Systems engineering focuses on interactions among project components, which may significantly amplify both benefits and costs. The analysis of risk and uncertainty in project planning illuminates the possible impacts of deviations from "best estimates" in projections of benefits and costs. The development of engineering methods of analysis for ecosystem restoration provides a way to integrate ecological components within more traditional Corps planning analysis and approaches.

10. The Corps should strengthen its programs in the areas of systems engineering aspects of water resources, risk and uncertainty analysis, and the integration of engineering and ecosystem analyses. Part of this strengthening should include the development of updated design manuals that better reflect contemporary methods and theories. These manuals should be used as general guidance rather than as "cookbooks" that specify a series of steps that must be strictly adhered to.

Independent Review

Water management theories, practices, and methods are constantly being tested and advanced in many different disciplines, in a variety of institutions, and in nations outside the United States. The resources required to employ experts that can stay abreast with these advances in all relevant disciplines and specialties, however, transcends the budgets of most federal agencies. Nonetheless, it is important that the Corps be familiar with current thinking and practices. An alternative to employing a suite of experts in every Corps District office is to periodically enlist independent experts to review and comment upon programs and on planning studies. Given the level of sophistication in Corps planning studies today, the participation of independent experts is often useful in ensuring that methods employed are consistent with current and credible thinking and practice. In today's planning environment, independent, expert advice is also essential for credibility. Other longer-term, potentially useful means for infusing ideas into the Corps are through the existing "visiting scholars" programs at the Corps Institute for Water Resources and the revived "Associates Program" within the Corps.

11. Independent experts from outside the Corps of Engineers should be routinely enlisted to provide advice in Corps programs and planning studies.

1

Corps of Engineers Missions, Projects, and Planning

INTRODUCTION

For nearly two centuries, the U.S. Army Corps of Engineers has planned and constructed much of the nation's civil works projects for navigation enhancement, flood damage reduction, and more recently, ecosystem restoration. As the nation's water resources needs and preferences have shifted, the Corps' mission has become more complex. For example, in the early twentieth century, the Corps focused largely on constructing channels and harbors, locks, and dams. By contrast, at the close of the twentieth century, the Corps was operating in a more complex legal context (including the 1969 National Environmental Policy Act, the 1973 Endangered Species Act, and various Clean Water Acts), was addressing a broader range of water-related needs such as recreation, and had assumed a new environmental restoration mission.

As the Corps' mission has evolved, the methods and techniques used in its project planning have become more sophisticated and more detailed. Corps of Engineers planning studies today often include economic models, long-range economic demand forecasts, and assessments of environmental impacts and benefits. In addition to analytical challenges posed by integrating economic, engineering, and environmental theories and methods, the Corps is striving to respond to input from stakeholder groups that may not be consistent with scientific principles or with the agency's planning procedures. Moreover, many Corps planning decisions today do not focus on a traditional emphasis of new civil works construction, but rather on operations of existing projects, which frequently entails the very different task of resolving resource trade-offs among competing interest groups.

Although its planning methods have often been subject to criticism, challenges to the Corps were especially pointed and prominent during the late 1990s and were perhaps best embodied by the close inspection of the Corps draft feasibility study of the Upper Mississippi River-Illinois Waterway (UMR-IWW). The economic analysis and modeling conducted by the Corps came under such intense scrutiny that early in 2000 the U.S. Department of Defense requested the National Academies' National Research Council (NRC)[1] to appoint a committee to review the economics components of the draft study (see NRC, 2001). In addition to challenges from outside the agency, the Corps is also in the midst of substantial internal changes, as the agency announced a major reorganization plan ("USACE 2012") in October 2003. The reorganization is intended to change the internal workings of Corps Headquarters and its regional elements and to promote a more business-like approach in order to streamline internal processes (*http://www.hq.usace.army.mil/stakeholders/*; accessed February 7, 2004).

Controversies and challenges surrounding the Corps' analytical techniques are far from resolved, however, and the U.S. Congress (along with many other groups and individuals) continues to debate the appropriate future vision of the nation's river and coastal systems and appropriate future roles for the Corps of Engineers. As part of this debate, in Section 216 of the Water Resources Development Act of 2000 (WRDA 2000), the U.S. Congress requested that the National Academies study the Corps' review procedures and its methods of analysis (Section 216 of the WRDA 2000 can be found in Appendix A).

In response to this congressional mandate, the Academies' National Research Council appointed four study panels and a coordinating committee. The first panel on review procedures completed its report in 2002 (NRC, 2002b). The other three panels assessed various aspects of Corps planning: project planning, river basin and coastal systems planning, and adaptive management. The coordinating committee tracked progress and promoted dialogue among the four study panels and also produced its own report. The studies were organized under the auspices of the NRC's Water Science and Technology Board, in collaboration with the Ocean Studies Board, and with input from the Board on Environmental Studies and Toxicology and the Transportation Research Board. The panels shared their thoughts and progress through different channels: chairs of each of the study panels were members of the coordinating committee; all panels and the coordinating committee were represented in a working session held in Irvine, California in November 2002; and panel and coordinating committee members attended meetings of panels of which they were not members. Each report from the panels

---

[1] The National Academies consists of the National Academy of Sciences, the National Academy of Engineering, and the Institute of Medicine. The National Research Council is the operating arm of The National Academies.

and the coordinating committee stands as an independent study (each of which was subjected to standard NRC review procedures), as well as part of this larger group of "216 studies." This is the report of the Panel on Methods and Techniques of Project Analysis, which was charged to examine and make recommendations for improving Corps planning methods for specific projects (as opposed to planning methods for river basins or coastal systems). Box 1-1 contains the panel's charge.

CORPS OF ENGINEERS PROJECT PLANNING METHODS AND GUIDANCE

Corps of Engineers planning studies are guided by many sources. The two most important are the 1983 federal Economic and Environmental *Principles and Guidelines for Water and Related Land Resources Implementation Studies* (also known as the *Principles and Guidelines*, or

BOX 1-1
Charge to the Panel on Methods of Project Analysis

This panel will review the Corps' project formulation and evaluation techniques and methods, including the use of models for project-specific applications (i.e., benefits, costs, optimization, etc.), with a view toward developing and refining planning methods to better serve the nation. As appropriate, the panel should also consider the methods, techniques, and practices employed by other federal agencies, states, and the private sector in the development of the projects.

The panel will also consider the need for system-wide considerations in project-specific planning and other state-of-the-art methods. The need to modernize and/or better implement the Corps methods and other techniques will be evaluated in the context of the current federal *Principles and Guidelines* (*P&G*). In addition, the panel will examine the interpretation of the *P&G* as reflected in Corps guidance such as ER [Engineering Regulation] 1105-2-100. This panel will also perform an ex post analysis of a sample of Corps projects (including major project purposes) based on 1) the methods used by the Corps and 2) state-of-the-art methods. As appropriate, the panel will make recommendations for improving Corps methods and techniques and may make recommendations regarding areas of the *P&G* in need of modernization.

simply *P&G*; WRC, 1983), and the Corps' Planning Guidance Notebook (PGN), known as Engineering Regulation (ER) 1105-2-100 (USACE, 2000). The two-page "principles" of the *P&G* aim to ensure "proper and consistent planning by Federal agencies in the formulation and evaluation of water and related land resources implementation studies" (WRC, 1983, p. iv). The "guidelines" consist of more than 100 pages and multiple appendixes of instructions for conducting planning studies for a range of water sectors (e.g., urban flooding, deep-draft navigation, recreation). The *P&G* replaced a planning document familiarly known as the "*Principles and Standards*," which was also approved by the federal Water Resources Council (WRC, 1973). A critical difference between the documents is that the *Principles and Standards* constituted planning **requirements**, whereas the *Principles and Guidelines* serves as **recommended guidance** with no legal force. In addition to the *P&G*, the Corps' *Planning Guidance Notebook* includes and builds on the *Principles and Guidelines* and provides additional guidance on implementing the *P&G*.

The Corps of Engineers conducts its planning studies in two stages: a reconnaissance stage and a feasibility stage. The reconnaissance stage is conducted to determine if a federal interest exists in a given water resources problem or opportunity. If a federal interest is determined to exist,[2] a feasibility study is then conducted. According to the *Principles and Guidelines*, feasibility studies must identify the national economic development (NED) alternative. This is the planning alternative that reasonably maximizes net national economic benefits, consistent with the federal objective of, as specified in the *P&G* principles, "protecting the Nation's environment, pursuant to national environmental statutes, applicable executive orders, and other Federal planning statutes" (WRC, 1983). Other alternatives may also be identified, but the NED alternative is the only alternative required by the *P&G*. The NED alternative may ultimately not be the alternative selected because a community or local sponsor may select a different plan. If a community implements a plan that goes beyond NED however, such as constructing a levee higher than the NED-recommended levee, that community is responsible for some or all of the additional costs.

Several Corps feasibility studies were reviewed during the course of this study, and several panel members spoke with Corps of Engineers planners and analysts. Sites of planning studies that were investigated

---

[2] The large majority of reconnaissance studies—approximately six of seven—conclude that a federal interest in the water resources issue at hand does not exist.

during the course of this study were Egg Harbor, New Jersey; Galveston, Texas; Oregon Inlet, North Carolina; St. Paul, Minnesota, and Wilmington, North Carolina. Observations gained in the review of these studies, plus panelists' knowledge of other planning studies, revealed both similarities and differences across Corps planning studies, fundamental assumptions, models employed, and Corps district offices. Given the variations in water resources problems across the nation, differences in approaches and techniques employed are to be expected. A limited degree of standardization within these detailed analytical methods, however, makes it difficult to draw agency-wide conclusions regarding Corps planning methods, models, or approaches. Moreover, the Corps of Engineers recognizes and uses few models universally in all of its district offices. More detailed investigations of this range of planning studies were beyond the resources and the time constraints of this study, but knowledge gained in these reviews was used to supplement and support the report's broader findings and recommendations, as appropriate, as they relate to the planning guidance within the *P&G* and the *PGN* (Chapter 3 also refers to other Corps projects and planning studies to help illustrate the broader issues raised therein).

CORPS OF ENGINEERS MISSIONS: HISTORICAL, CONTEMPORARY, FUTURE

The Corps of Engineers' traditional missions have been to reduce flood damage and to enhance navigation on the nation's inland waterways and in its ports. These missions were often carried out via the civil engineering activities and structures for which the Corps is well known: damming of rivers and creation of navigation pools and storage reservoirs, construction of dikes and levees in floodplain and other low-lying areas, beach protection projects in coastal areas, and the dredging of ports and rivers. These projects and activities are still requested by the nation's citizens and elected leaders, and they remain an important part of the Corps' work program. The extent of the Corps' involvement in these traditional areas, however, has diminished for several reasons: most of the best dam sites have been dammed; concerns regarding environmental impacts of these projects; a substantial decrease in the Corps' civil works budget since the mid-1960s (more than 50 percent; USACE, 2001a); and an increasing demand for nontraditional activities, primarily ecosystem restoration.

The trend of the Corps building fewer new civil works projects is likely to continue. This is not to say that the Corps' importance to national water management will diminish. The Corps will retain responsibility for maintaining and operating a massive, multibillion dollar infrastructure that controls hydrologic and related processes in many of the nation's river and coastal systems (e.g., the Missouri River dam and reservoir system). Another relevant trend is that stakeholder groups today request a greater voice in decision making than in the past. Interest groups and citizens alike are today also less likely to trust the Corps' professional judgment and analyses than in the past. Corps of Engineers planning studies today often attract much interest and may be carefully scrutinized. The Corps is aware of this and typically hosts public meetings and other events to solicit public input on the agency's highly visible planning studies. Changes in budgetary constraints and congressional priorities, changes in philosophies of successive administrations, shifts in social preferences, and changing planning paradigms have converged to create a dynamic setting for the Corps' civil works program.

REPORT OVERVIEW

Following this introductory chapter, Chapter 2 provides a review of the structure and evolution of federal water management organizations and policies. Chapter 3 focuses on economic theories and approaches within the *Principles and Guidelines*. Chapter 4 discusses the Corps' efforts to promote public participation and input in planning studies. Chapter 5 reviews the engineering models used by the Corps. Chapter 6 discusses and provides the panel's recommendations; in doing so, it draws on previous chapters of this report. Chapter 7 is the concluding chapter and provides a brief epilogue of Corps of Engineers planning studies, guidance, and methods.

2

Federal Water Resources Planning Objectives and Guidance

This chapter reviews the historical basis of the federal government's interest in water resources development. It examines federal laws, policies, and guidance that influence current actions of the Corps in planning water resource projects, and processes and procedures used by the Corps to carry out its mandates. To provide a better understanding of Corps planning, it also describes those organizations and groups that participate in the development of a Corps project. The chapter concludes with some perspectives regarding Corps planning guidance, the planning process for federal water projects, and the current national water policy and organizational landscape.

FEDERAL INTERESTS IN WATER RESOURCES

Origins of Federal Involvement

In the late eighteenth century, water-related problems were the responsibility of state and local governments. As the magnitude of these problems was recognized, federal roles soon emerged. As early as 1824, the Congress envisioned a federal role in addressing water problems and issues when it authorized the Corps to investigate navigation potential on the Ohio and Mississippi Rivers. Interest in alleviating flood problems led to the federal Swamp Lands Acts of 1849 and 1850. In establishing the Mississippi River Commission in 1879, Congress placed the Corps of Engineers in charge of navigation on the Lower Mississippi and gave it primarily advisory responsibilities for assisting with flood problems along the entire Mississippi River. The 1902 Reclamation Act established the U.S. Reclamation Service (renamed the Bureau of Reclamation in 1923) and sought to irrigate arid lands and increase settlement in the arid regions of the western United States. This was followed in 1917 by federal assumption of increased responsibilities for managing floods on the Sacramento and Lower Mississippi Rivers. Shortly thereafter, Congress exercised its powers to establish standards for non-federal development of hydroelectric power on navigable waterways. In 1933, President Franklin Roosevelt established the federal Tennessee Valley Authority to manage water and related natural resources on a basin-wide scale. Following devastating floods across the country in the late 1920s and early 1930s, Congress passed the Flood Control Act of 1936, which made flood management a federal responsibility. This act also enshrined the practice of weighing a proposed project's costs against its benefits, in effect subjecting all future Corps of Engineers planning studies to a benefit-cost test. In 1944, Congress passed the Pick-Sloan Plan, which represented a merger of the Corps of Engineers' "Pick Plan," which was designed primarily to manage floods on the mainstem Missouri River, and the Bureau of Reclamation's "Sloan Plan," which was designed primarily for irrigation. The Pick-Sloan Plan called for the Corps to construct mainstem reservoirs for flood control and navigation enhancement, with the Bureau of Reclamation constructing tributary reservoirs for irrigation and hydroelectric power production. In the two decades following World War II, Congress expanded federal responsibilities in water resources to authorize the Corps to include recreation, hurricane protection, beach protection, and water supply in its projects. In the 1990s, Congress broadened the Corps' responsibilities to include environmental restoration as a primary Corps project output. In summary, the period from 1850 to 2000 saw the federal role in water resources expand from minimal programs and responsibilities to a central role in water resources project construction, management, and operations, as well as in science-related programs.

Emergence of Economics-Based Frameworks

Decisions about water resources projects and other public works became more systematic in the late nineteenth century. Despite increased employment of professionals within the federal government and an increasing reliance on experts, Congress was reluctant to cede power regarding water project selection. Although demands for fiscal accountability were limited in this era, advocates for greater efficiency worked to reduce the number of projects that were primarily based on political considerations. In 1902, for example, a Board of Engineers for Rivers and Harbors (BERH) within the Corps was established and direct to certify that projects were beneficial before they could be recommended to Congress (Porter, 1996). By the 1920s, the BERH expected the benefits of projects it recommended to exceed the costs. With a request from Congress in 1927 to study all major river basins in the United States (the "308 reports"), the Corps civilian work force began to increase and the agency increasingly relied on quantitative methods in its studies to impose discipline (Porter, 1996).

The 1936 Flood Control Act is best known for its requirement that " . . . the Federal Government should improve or participate in improvements . . . for flood control purposes if the benefits to whomsoever they accrue are in excess of the estimated costs." This requirement remains in place today and has been a fundamental principle in Corps planning studies over the decades (Chapter 3 of this report examines the use of cost-benefit analysis in Corps planning studies). A 1941 National Resources Planning Board report identified the need for economic analysis in evaluating water projects (but also highlighted the need to consider intangible benefits of the projects; Holmes, 1972). In 1952, the Bureau of the Budget (now the Office of Management and Budget) issued detailed instructions on the form and conduct of such project analyses. These efforts were " . . . economics centric and established the economic benefit-cost ratio as the principal evaluation tool."

Principles, Standards, and Guidelines

The increased focus on economics in water resources project analysis drew the attention of both the Congress and presidential candidate John Kennedy, whose platform included broadening the criteria for water resources project evaluation to include environmental and social benefits. Kennedy's efforts, as well as those of Congress, resulted in passage of the Water Resources Planning Act in 1965, which culminated decades of efforts toward more centralized water resources planning. The 1965 act had two components: Title I, which established a federal Water Resources Council (WRC), and Title II, which provided the framework for establishing interagency-interstate commissions. The Water Resources Council initially consisted of seven cabinet-level departments: Agriculture, Army, Commerce, Energy, Housing and Urban Development, Interior, and Transportation. The Environmental Protection Agency (EPA) was added in 1970.

The Water Resources Council was directed to establish principles and standards to guide justification of federal water projects. In the late

1960s, a WRC Task Force proposed principles and standards that established four accounts against which projects developed by the Corps of Engineers, the Tennessee Valley Authority, the Bureau of Reclamation, and the Natural Resources Conservation Service (then, the Soil Conservation Service) should be evaluated. The four accounts were national economic development, environmental quality, regional economic development, and social well-being. The Flood Control Act of 1970 included congressional "intent" that the four objectives be equal. In 1973, the *Principles and Standards for Planning Water and Related Land Resources (P&S)* document was published in the *Federal Register* (WRC, 1973). The *P&S* established environmental quality (EQ) and national economic development (NED) as coequal objectives. These two accounts were to be displayed in project justification along with information on regional economic development and social well-being. Before much experience could be gained with the WRC's *Principles and Standards*, President Carter took office and modified the *P&S* to emphasize the equality of the two principal accounts (EQ and NED). One year after taking office, President Reagan rescinded the *P&S* and issued in its place the *Principles and Guidelines for Water and Related Land Resources Implementation Studies* (*P&G;* WRC, 1983). The *P&G* document established the single objective of federal water resources development as "to contribute to national economic development consistent with protecting the nation's environment." Project analyses could include information on the social and regional accounts, but this information had little influence on planning decisions. Since 1983, the *P&G* has remained the key planning guidance for federal water projects (of relevant agencies) despite criticisms of the document's narrow focus and its failure to adequately incorporate nonquantifiable environmental and social impacts into its planning steps.

Environmental Legislation

The 1965 Water Resources Planning Act brought increased attention to environmental considerations. Additional attention to environmental issues was stimulated by passage of the National Environmental Policy Act (1969), the Endangered Species Act (1973), the Clean Water Act (1972, and amended in 1977), and related legislation. These laws significantly modified the process for considering environmental issues in federal water projects. Environmental impact analyses had to be conducted and submitted with project plans. Those plans had to indicate compliance with the other acts and close coordination with relevant agencies, especially the U.S. Fish and Wildlife Service. By the end of the 1970s, environmental concerns were firmly established within federal water resources project planning.

Since the mid-1980s, Congress has provided considerable legislative direction to the Corps concerning environmental issues in its (roughly) biennial Water Resources Development Acts (WRDA). Examples include the 1986 WRDA, which directed fish habitat improvements, authorized environmental studies to help adjust operations of completed projects (also see Chapter 6), and authorized changes in justification for beach nourishment projects. The 1990 WRDA directed the Secretary of the Army to include environmental protection as a primary Corps mission. The 1996 WRDA authorized the Secretary of the Army to carry out aquatic ecosystem restoration and protection projects and to add environmental protection and restoration as another project purpose (P.L. 104-103). These authorizations, however, have rarely been accompanied by adequate financial support in administration budgets or in funding from congressional appropriations committees.

Cost-Sharing

Until 1986, the bulk of the construction costs of Corps water projects was borne by the federal government, although some forms of cost-sharing existed since the early twentieth century. In 1986, believing that local governments should play a greater role in water project development, and in seeking local government approval of federally-sponsored projects, President Reagan submitted a proposal to Congress calling for changes to cost-sharing provisions for all Corps projects, with the proposed shares of local cost-sharing responsibilities varying by project purpose. After considerable debate in Congress, the 1986 Water Resources Development Act was passed and signed into law. The cost-sharing arrangements of WRDA 86 resulted in significant changes to then-existing cost-sharing formulas (see Table 2-1). The enactment of more stringent cost-sharing requirements resulted in local sponsors' understandably requesting a stronger voice in plan formulation. Although a sound principle in many ways, cost-sharing has resulted in the Corps often experiencing tensions between local sponsors calling for a locally-preferred alternative, and the Corps' obligation to uphold the public interest (NRC, 1999a).

TABLE 2-1 Project Cost-Sharing Formulas for Corps of Engineers Projects

Purpose	Non-federal Share
Navigation, harbors	20%: depth < 20 ft. 35%: depth 21-45 ft. 60%: depth > 45 ft.
Navigation, inland	50% from fuel tax
Flood control	
Structural	min. 35%–max. 50%
Nonstructural	25%
Hydroelectric power	100%
M&I water supply	100%
Agricultural water supply	35%
Recreation	
Navigation	50%
Other	50% of separable cost
Hurricane and storm damage	35%
Aquatic plant control	50%

SOURCE: Adapted from USACE (2002c).

CONTEMPORARY PROJECT PLANNING AND GUIDANCE[1]

The Study Initiation, Review, and Approval Process

Many water resources planning studies begin at the local level with a perceived need or opportunity. This need is typically discussed with the responsible Corps of Engineers district office, which advises the interested parties on actions needed to move their request forward. The given local interest group or community typically requests assistance from its congressional delegation. Through congressional channels, a request is forwarded to the Corps to study the feasibility of a project and to determine if it represents a federal interest. If congressional support exists, the Corps is directed through a resolution from either the Senate or the House authorizing committee or in a Water Resources Development Act,

[1] This document addresses the current planning process of the Corps and the documents that support that process. Because many Corps projects are several decades old, documents associated with them may have different names than those discussed in this report (e.g. General Design Memorandum). Their purposes were similar, if not the same, as many present documents.

to examine the need and report to Congress on the feasibility of a project. For example, a request for the Corps to investigate expansion of navigation on the Great Lakes was part of the Water Resources Development Act of 1999 (P. L. 106-53, Sec. 456, 113 Stat. 269, 1999) and read:

> In consultation with the St. Lawrence Seaway Development Corporation, the Secretary shall review the Great Lakes Connecting Channels and Harbors Report dated March 1985 to determine the feasibility of undertaking any modification of the recommendations made in the report to improve commercial navigation on the Great Lakes navigation system, including locks, dams, harbors, ports, channels, and other related features.

Reconnaissance and Feasibility Studies

Before any study can take place, Congress must appropriate funds for the study. If funds are appropriated, the Corps conducts a reconnaissance study to determine the feasibility of a project, to identify a local sponsor for the project, and most importantly, to determine whether a federal interest exists. A reconnaissance study currently is limited to one year and is to cost no more than $100,000. It does not normally involve substantial public input. Reconnaissance studies are conducted by Corps district offices. If a federal interest is identified and if Corps Headquarters approves the study, a district office proceeds to the feasibility stage (as noted in Chapter 1, the large majority of reconnaissance studies do not lead to a feasibility studies).

Initiation of a feasibility study requires the administration to approve a budget request by the Corps to conduct the study, the appropriation by Congress of funds to support the feasibility study (costs can range from one million to several million dollars), and an agreement by an approved local sponsor to pay half the costs of the feasibility study. Conduct of a feasibility study ideally takes about three years, but often requires up to five years or more. A 1999 National Research Council committee noted that feasibility studies on average require about 4.5 years to complete (NRC, 1999a). During the study period, the administration must include study funds in the president's budget and Congress must annually appropriate study funds. The feasibility study also involves preparation of environmental impact statements, significant public involvement, and coordination with state and local officials.

As noted earlier, the Corps uses several guidance documents in its water resources planning studies. The *P&G* and the *Planning Guidance Notebook* have been mentioned. Other important documents include the Corps *Digest of Water Resources Policies and Authorities,* more than 60 planning guidance letters, 23 economic guidance memoranda, and a series of engineering regulations (ERs) and engineering circulars (ECs; these documents are listed at *http://www.usace.army.mil/inet/functions/cw/cecwp/General_guidance/guidance.htm*).

The planning process described in the *Principles and Guidelines* includes six steps:

1. Specify problems and opportunities.
2. Inventory and forecast conditions.
3. Formulate alternative plans.
4. Evaluate effects of alternative plans.
5. Compare alternative plans.
6. Select recommended plan.

These steps are not intended to be strictly sequential, and there may be iterative feedback among steps as more information is gathered and analyzed during a planning study. The Corps usually begins steps 3 and 4 within the first several months of the feasibility study, but the timing of these activities depends on the scale and complexity of the problem or issue at hand. At this stage, the Corps conducts project design analyses (including engineering and hydrologic studies) and estimates project benefits and costs. When a Corps district office is prepared to present the alternative plans, an alternative formulation briefing (AFB) is usually conducted.

This alternative formulation briefing is held to facilitate early Washington-level acceptance of the plan formulation and selection process, of the identified preferred plan, and of proposed federal and non-federal responsibilities. The goal of this briefing is to allow a Corps district office to release a draft report to the public concurrent with Washington-level policy compliance review of the report (see USACE, 2001). Local sponsors and other interested parties, including technical experts, participate, and the public is invited.

After the alternative formulation briefing, the Corps district office prepares the draft feasibility report and a related environmental impact statement (which undergoes public review under the parallel National Environmental Policy Act process). When the feasibility study is completed at the district level, and after coordination with other federal agencies and state officials, it is submitted (via a Corps division office) to Corps Headquarters for approval. The final step in the formal planning process is approval of the final feasibility study by the Chief of Engineers. This approval is in the form of a short letter to the Assistant Secretary of the Army (Civil Works) (ASA-CW). Following reviews by the ASA-CW and by the Office of Management and Budget, and with the concurrence of the administration, the study identifies a proposed project. It is then transmitted by the administration to Congress for authorization and for appropriation of funds for implementation.[2]

All projects forwarded to Congress by the administration must be reviewed and authorized by committees of both Houses. Such action follows testimony from local supporters and opponents and discussion of the project by the Corps. Approximately every two years, Congress prepares a WRDA, part of which authorizes construction of projects by the Corps and authorizes, but does not appropriate, funds for these projects. Project plans may proceed only with this congressional authorization.

Summary

The time required for a project to move from conception to completion depends on its size and complexity. Factors that affect cost and time requirements include controversies surrounding the project, actions of the administration, Congress, and state and local officials, and the ability of local sponsors to provide their share of the project costs. About 20 years ago, the average project required 25 years to move through this process. In comparison, a non-controversial project today averages 7 to 8 years, with the average of all projects being 15 years (Fred Caver, U.S. Army Corps of Engineers, personal communication, 2002). This shortening of the project life cycle has resulted from changes in the planning process, including increased public participation and a reduction in the number of required project reviews. Controversial projects and those without substantial support, however, still require decades to move through the conception-completion cycle, or to reach a point at which they are finally de-authorized. Many projects do not make it through the process and are dropped along the way because they encounter problems, are not justified, are not in the federal interest, or lack local support. At the end of 2002, the Corps reported that only 33 percent of the 543 pro-

---

[2] Congress has been known to bypass the administration and not wait for its (the administration's) submission.

jects authorized for construction were actually funded in the fiscal year 2003 budget (Table 2-2). Seventy projects lacked either a sponsor or local support and were inactive.

PARTICIPANTS IN WATER RESOURCES DECISION AND POLICY MAKING

The Administration

Although the Corps of Engineers is a key agency in the federal water project planning and construction process, Congress, the administration, state and local officials, the public, and local sponsors all play significant roles. Congress and the administration have sparred for decades for primacy in the process. President Carter began his administration with a water project review and established new priorities for water project approval. Congress largely stonewalled his efforts. President Reagan made significant changes in water policy rules and made many of them stick, and established a cabinet council on natural resources to develop policy. Much of President Clinton's water policy was based on initiatives within the Office of the Vice President and the chair of the Council on Environmental Quality.

The federal Water Resources Council represents an interesting part of the description of administration leadership. It is an interagency group of seven cabinet officers, along with the EPA administrator, who share water-related responsibilities. Established in the 1965 Water Resources Planning Act, the WRC's principal duties are to (1) assess the adequacy of regional and national water supplies; (2) study regional or river basin plans and programs in relation to larger national regions; (3) assess the adequacy of administrative and statutory means for the coordination of water and related land resources policies and programs of federal agencies; (4) appraise the adequacy of existing and proposed policies and programs to meet these requirements; and (5) make recommendations to the president with respect to federal policies and programs. Although still legislatively authorized (42 U.S.C. Sec. 1962 a-1 to a-4), the WRC has not been funded since 1981. Funding was discontinued in large part because of problems related to the expansive and divergent duties with which it was assigned. It was criticized for doing too little to confront difficult policy issues and interagency conflicts, and for getting too involved in review of individual projects of the basin commissions. Be-

TABLE 2–2 Funding Status of Authorized Corps Construction Projects (million dollars)

	Number of Projects	Projects with FY 2003 Funding	Total Cost	Total Allocated to Date	Funded Amount (FY 2003)	Cost to Complete
Total projects authorized for construction	543	179	$62,663	$24,571	$1,373	$36,717
Active projects	456	179	$60,063	$24, 056	$1,373	$34,633
Deferred – sponsored or politically supported projects	17	0	$343	$103	$0	$239
Inactive non-sponsored or non-politically supported projects	70	0	$2, 257	$412	$0	$1,845

SOURCE: USACE (2003).

cause much of its staff was on detail from federal agencies or had served on agency staff, and because meetings of the council members or their representatives involved high level members of the administration, the WRC staff became a clearinghouse for new ideas as well as problem identification. In 1973, the National Water Commission concluded, "[t]he Council's potential for leadership in policymaking and in planning activities has not been realized" (NWC, 1973). Despite these alleged failings and criticisms, many experts and study commissions have promoted reinstituting the organization in order to coordinate the growing number of water-related programs and policies.

Office of Management and Budget

In every administration, the Office of Management and Budget (OMB) has played a significant role in prioritizing water resources projects. The OMB provides the White House with alternative budget approaches and identifies what might be feasible within fiscal guidelines sought by the president. The OMB translates broad White House guidance into instructions to the Corps (and all federal agencies). It establishes priorities for funding and approves new project starts. The OMB reviews all projects submitted to Congress for compliance with administration policy and, when appropriate, returns them to the Corps for revision. OMB approves all statements made by members of the administration to Congress dealing with policy or funding. OMB is "the eye of the needle" through which federal water resource projects must pass.[3]

Council on Environmental Quality

Established pursuant to the National Environmental Policy Act in 1970, the Council on Environmental Quality (CEQ) has played various roles in water resources development. Throughout the period, the constant has been its statutory responsibility to oversee the NEPA process and to serve as the administration's voice on contentious environmental impact statements. If another federal agency objects to a project proposed by the Corps, the CEQ serves as an arbiter. The CEQ also reviews projects when the project justification is based on substantial environmental components, such as wetland restoration. The CEQ's role within federal water development has shifted over time in response to the status of the CEQ chair in the White House.

U.S. Congress

Each new biennial session of Congress brings changes in terms of committee structure, congressional leadership, and priorities. As a result, Congress' relationship with Corps activities has varied over time. Individual members can become critics or advocates of Corps activities and, depending on their seniority or leadership position, can significantly influence such activities. However, the most influential components of Congress are the committees charged with authorizing national water policies and Corps projects and programs and the committees that annually appropriate funds for projects and studies.

Both the U.S. Senate and the U.S. House of Representatives are organized with committees for authorizing projects and committees for appropriating funds for these projects. In the Senate, the Committee on Environment and Public Works and its Subcommittee on Transportation, Infrastructure and Nuclear Safety and, in the House, the Committee on

---

[3] Congress has frequently questioned the OMB's review role.

Transportation and Infrastructure and its subcommittee on Water Resources and Environment deal with authorizations. Appropriations committees exist in both bodies and operate though the Subcommittee on Energy and Water Development in both the Senate and the House. These four committees are critical to the Corps civil works program but are not the only ones dealing with water and Corps operations. More than 10 congressional committees have responsibility for aspects of water resources development.

As with all congressional committees, considerable power is vested in the committee chair and the ranking member (the latter represents the minority party), and with subcommittee chairs and ranking members. In turn, most program evaluation and analysis is handled by professional staff, many of whom have served with the committees for years.

Assistant Secretary of the Army for Civil Works

Most legislation directing the Corps to construct water projects actually directs this task to the Secretary of the Army. To oversee civil works activities, in the Flood Control Act of 1970, Congress established the position of Assistant Secretary of the Army for Civil Works (ASA-CW), a political position requiring presidential appointment and Senate confirmation. The Assistant Secretary of the Army for Civil Works is charged with policy oversight and general management of civil works duties, as well other U.S. Army civil works responsibilities, and is assisted in carrying out his or her duties by a small staff. The ASA-CW serves as the principal liaison between the White House (including OMB), Congress, and the Army for civil works matters. The ASA-CW is responsible for reviewing all projects submitted to OMB for compliance with administration policy, and thus often becomes the focal point for discussions between members of Congress, project supporters, and the administration on controversial projects.

Contrary to popular perception, the Assistant Secretary of the Army for Civil Works is not the head of the Corps of Engineers. A military officer selected by the president and confirmed by the Senate holds the position of Chief of Engineers and Commander, U.S. Army Corps of Engineers. The commander operates under the policy guidance of the ASA-CW for civil works matters and the policy guidance of other assistant secretaries for other aspects of the activities of the Corps (military construction, research and development, and military operations).

The Corps of Engineers

Organizational Structure

The U.S. Army Corps of Engineers is a major command of the U.S. Army. It is the world's largest public engineering, design, and construction management agency, with a variety of missions in the U.S. and overseas. These include responsibility for military construction in support of the Army at bases worldwide, support of select Air Force construction, and research and development in engineering. The Chief of Engineers also serves as the principal engineering adviser to the Chief of Staff of the Army. In the civil works area, the Corps carries out this function in all 50 states and in the Pacific islands (Box 2-1 explains why water resources projects and other civil works functions are under the aegis of the U.S. Army).

The Corps is organized into a headquarters in Washington, D.C.; eight Divisions (normally headed by a general officer), and 41 districts (commanded by colonels and lieutenant colonels of the regular Army; see Figure 2-1 and Appendix B). Corps districts and divisions are distributed across the U.S., Europe, and the Pacific. Within the U.S., the Corps' civil works organizational boundaries normally represent all or part of river basins.[4] Although military officers command the divisions and districts, essentially all staff consists of career members of the civil service. As of November 2002, there were 450 military personnel and 34,707 civilian personnel assigned to the divisions and districts. The largest district had 1,362 personnel; the smallest, 157 (Susan Duncan, U.S. Army Corps of Engineers Director of Human Resources, personal communication, December 30, 2002).

Districts are the Corps' operating units. Some districts have both civil works and military functions, while others have only military or only civil works functions. Although there is considerable variation in size and capability among district offices, most have regulatory, planning, engineering design, construction, and operations functions. Operations activities include the operations of locks and dams, reservoirs, and other facilities constructed by the Corps. Districts operate under regulations and guidance documents promulgated by the Chief of Engineers, but differences among districts often exist in the implementation and interpretation of guidance documents.

---

[4] Regulatory functions are by state.

BOX 2-1
Civil Works Functions in the Army

The Army Corps of Engineers has been in the middle of national public works activities for nearly two centuries. In the late eighteenth century, to the military was turned to for assistance in developing ports, harbors, and rivers. In 1802, the U.S. Military Academy at West Point became the country's first engineering school, and remained its only engineering school for more than three decades. As settlement expanded westward, Army engineers led the exploration, helping to build roads and railroads. In 1824, Congress assigned the Corps responsibility for surveying and, eventually, for the development of inland navigation. Over the next 150 years, Congress assigned the Corps with responsibilities in controlling floods, generating hydropower, and providing recreation sites. Congress also assigned the Corps responsibility for regulating activities in navigable waterways and, more recently, activities in wetlands. These civil works functions were matched with the Corps' responsibilities for providing military construction support to the Army and the Air Force in peace and in war, as well as supporting the activities of other agencies. Synergy between civil works and military construction has resulted in the Corps helping to construct the Panama Canal, restoring European ports following World War II, constructing Cold War facilities around the world, providing assistance in Kuwait following the Gulf War, and assisting peacekeepers in Bosnia and Kosovo. Most recently, Corps civil works personnel were called to Iraq to assist in restoring some essential services and a portion of Iraq's physical infrastructure. Historically, the Defense Department has supported the Corps' continuation in civil works for the following reasons:

- The nationwide civil works organization of the Chief of Engineers provides an existing organization to support rapid mobilization by the nation's armed forces prior to or in the event of war.
- This civil works experience provides a means for training engineering leaders in complex, large-scale types of construction and in related logistics efforts encountered in wartime.
- Having a civil works organization to depend on provides economies and efficiencies for the Army and Air Force military construction programs.
- The civil works activities of the Army reflect favorably on the Army and enhance the Army's image at the grassroots level.
- The civil works organization provides experienced natural disaster recovery capabilities.

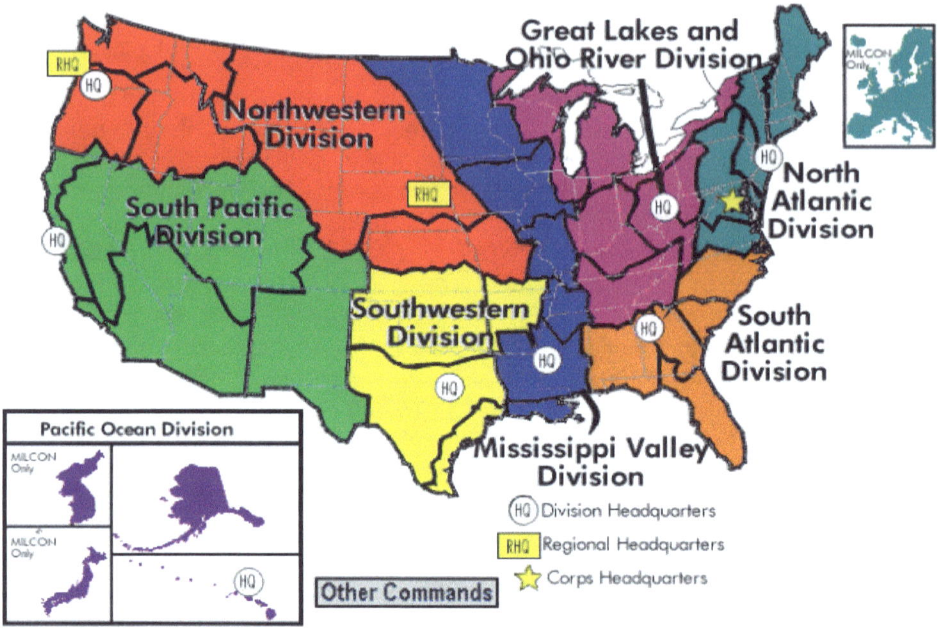

FIGURE 2-1 Corps of Engineers divisions and headquarters in the United States, Pacific, and Europe. SOURCE: Available on-line at *http://wwww.usace.army.mil/divdistmap.html,* accessed October 7, 2003.

The Corps and Local Citizens

Corps districts often develop strong relationships with the people that reside in their geographic region or whose organizational activities take place within the districts. Corps civilian staff frequently spend their careers in the same district or division and are members of communities or regions facing water problems. If there are local or regional water-related problems, citizens often turn to the Corps for advice. In addition, the Corps may already be working with local leaders and elected officials on existing projects. Some critics believe that this relationship can create a bias within the Corps; others see this relationship as enhancing responsiveness to local problems.

Other Federal Agencies

In executing its water resources development activities, the Corps must work closely with many federal agencies and departments. Its studies and projects must be coordinated with the Fish and Wildlife Service or with NOAA (the National Oceanic and Atmospheric Administration) Fisheries. The Environmental Protection Agency reviews the Corps' environmental impact statements. In the 17 western states, the Corps coordinates appropriate activities with the Bureau of Reclamation. Its hydropower generation activities are coordinated with the Department of Energy. Ports, harbors, and inland waterways involve close work with the Coast Guard and the Department of Transportation. Flood risk management projects must be tied to upstream small projects of the Natural Resources Conservation Service and to mitigation efforts of the Federal Emergency Management Agency. Some programs may require coordination with additional federal agencies and departments.

Interagency Coordination

Interagency coordination on water resource project development and water resources policy has taken several forms. During the 1930s, various boards established by President Franklin D. Roosevelt operated with water resources elements representing the principal federal agencies with water resources management responsibilities. With the abolition of these boards, the federal agencies established (without any statutory authority) the Federal Interagency River Basin Committee which, in turn, established regional interagency committees to foster basin-wide coordination. The Water Resources Planning Act of 1965 directed the establishment of the Water Resources Council, headed by the Secretary of the Interior with membership of the secretaries of the water resources agencies and the major river basin commissions. In 1981, President Reagan eliminated funding for the WRC and basin commissions.[5] Since the early 1980s, there has, thus, in effect been no federal coordinating body. For specific purposes, ad hoc committees have been established to address issues such as the 1993 Mississippi River floods. The Corps has, however, cooperated with other agencies in a "Federal Principals Group" designed to foster interagency communication in connection with the

[5] The WRC remains an entity within the federal government under the 1965 Planning Act. Because it receives no funding for personnel or activity, it neither convenes meetings nor issues reports.

Corps' ongoing Upper Mississippi River–Illinois Waterway feasibility study. A similar interagency group is also meeting to oversee wetland restoration activities in coastal Louisiana.

State and Local Officials, Project Sponsors, and the Public

The evaluation and construction of water projects requires support of state and local officials as well as an officially designated local sponsor. Typically, projects being considered at the local level are coordinated and supported by state governments. Opposition to a project by the legislature or governor of a state would be noted by the Corps in its planning documents and would inevitably lead the Corps to avoid requesting funds for the project.[6] Local governments frequently serve as project sponsors, responsible for providing the non-federal share of the costs. In other cases, they support the project sponsors fiscally and with in-kind support (engineering services, data, etc.). Depending on the circumstances, other governmental entities, such as conservation districts and levee boards with an ability to raise funds for the non-federal share of costs, serve as project sponsors.

The public, as represented by both interest groups and the citizenry at large, plays an important role in the project development process. At various stages in the movement of a project from conception to completion, the Corps conducts public meetings to obtain local views on the project. Throughout project development, interested parties also provide written comments on aspects of the project to the Corps, state and local officials, and members of Congress and the administration. Projects of regional or national importance draw the attention of national groups. As projects progress from stage to stage, the Corps typically issues press releases that announce its actions on the project and solicit comments.

Public comments are addressed formally in the environmental impact process and are noted in project documents. Controversial projects typically attract comments supporting and opposing the project, and the Corps reflects this diversity in its reports, but the Corps ultimately makes its recommendations based on all of the information generated in the study effort. In turn, the administration and Congress must consider the facts of the project report as well as the public comments.

---

[6] When he was Governor of Georgia, Jimmy Carter objected to construction of Sprewell Bluff Dam. Believing that the Corps had gone ahead with the project over his objections, he raised the issue when he became president, only to learn that the Corps had withdrawn all support once it learned of his opposition (Baldwin, 2000).

COMMENTARY

The Corps conducts its planning studies in a setting that consists of several layers of government and many other federal agencies, a large body of federal legislation and administration guidance, cosponsors and interest groups, and federal and Corps-specific planning guidance. There is a long history of congressional influence on the Corps program and projects, and the Corps, Congress, and local water project supporters have historically formed a system of mutually reinforcing support (a so-called "Iron Triangle"). These structures have perpetuated the water project authorization-appropriation-construction process over the decades and have proven resilient to challenges from various administrations. Nonetheless, the administration can exert substantial influence on the Corps through executive-level bodies, especially the OMB. Corps projects and planning studies have long been, and continue to be, subjected to influences in the political arena.

The Corps also conducts its studies with input and in relation to other federal agencies. There was a period in which the roles and responsibilities of the Corps and other water-related agencies were relatively clear. Over time, however, and with the accumulation of federal legislation and other directives, lines of authority between the nation's water resources agencies have become blurred. In an earlier era, for example, the Corps paid limited attention to aquatic habitat, leaving such concerns to the Fish and Wildlife Service and to state fish and game agencies. Today, however, environmental implications of the Corps' program are paramount; there are concerns about environmental impacts of Corps projects, interest groups spar with one another to gain a greater portion of benefits attached to Corps projects (e.g., river flows), and the Corps has been tasked by Congress to construct ecosystem restoration projects. Accordingly, the Corps today conducts its own ecological analyses and employs scores of ecological and biological scientists. The Corps finds itself bumping up against agencies such as the Fish and Wildlife Service and the U.S. Geological Survey in studies of the effects of its projects on aquatic habitat. The Corps is also obliged to abide by an increasingly complicated body of federal legislation.

Limited coordination between federal water-related agencies is not a new problem or observation. The President's Water Resources Policy Commission (1950), chaired by Morris Cooke, noted that "there is today no single, uniform Federal policy governing comprehensive development of water and land resources." The commission's report went on to say, "This Commission is therefore recommending the achievement of the necessary coordination through the unification of policy governing the actions of existing agencies, or of a single agency should such be adopted. This unification of policy should be assured through enactment of a single national water resources policy law . . ." (President's Water Resources Policy Commission, 1950).

More recently, a report from the National Commission for the Public Service (NCPS, 2003) reviewed the federal government's structure and explained the phenomenon of expanding agency missions and increasing sophistication and demands:

> In this technological age, the government's widening span of interests inevitably leads to complications as organizations need to coordinate policy implementation. But as things stand, it takes too long to get even the clearest policies implemented. There are too many decisionmakers, too much central clearance, too many bases to touch, and too many overseers with conflicting agendas . . . .The system has evolved not by plan or considered analysis but by accretion over time, politically inspired tinkering, and neglect . . . governmental reorganization has come to be viewed as a task so daunting, requiring such extensive and excruciating political negotiations, that it takes a national emergency to bring it about.

This quote explains well the current situation in federal water policy, as the Corps and several other agencies with water-related responsibilities conduct their respective programs and duties without a high-level body to ensure coordination, efficiency, and clear articulation of lines of authority. Since the Water Resources Council was zero-funded in the early 1980s, administrations have chosen to promote federal water-related programs without a formal coordinating body. Over this period, many analysts and committees, including a previous National Research Council committee that reviewed Corps planning (NRC, 1999a), have recommended the establishment of a federal water coordinating body (or a reinvigorated Water Resources Council). Before then, the value of a body to ensure interagency commission was recognized.

The National Water Commission Act of 1968 established a National Water Commission of seven distinguished nongovernmental members. In its 1973 report, the commission identified, in seven thematic areas, water resources issues that were likely to arise as the nation developed its water policy (NWC, 1973).[7] The commission analyzed these issues and offered recommendations for action. The commission also recommended strengthening the Water Resources Council by placing it in the Executive Office of the President with an independent full-time chairperson who would also serve other White House staff as a presidential adviser on water resource matters. By making the chairperson independent and with the status of a presidential adviser, the commission indicated that the WRC would have a broader point of view and the chairperson would speak for the president, rather than for one department, in interdepartmental conflicts and controversies. It also noted that other departments and agencies with water-related responsibilities could be included and that with these reforms, land use planning and water resources planning could be better integrated. The proposal received little support. Since the Water Resources Council was zero-funded in the early 1980s, there has been no progress toward invigorating or creating an executive-level body to promote water policy and interagency coordination. During this period, the Corps has found itself in the middle of more water resources controversies that it finds difficult, if not impossible, to successfully resolve.

In a context of conflicting legislative and other directives, the methods by which water resources projects are evaluated and selected becomes paramount in allocating federal funds. The guidance in the *Principles and Guidelines* allows project evaluation procedures to vary somewhat, according to interpretations by individual district offices. Chapter 3 examines the methods used to evaluate economic and environmental approaches of candidate projects.

---

[7] The thematic areas were water demand, the shift from water development to restoration and enhancement of water quality, the tie between land and water planning, water conservation, economic principles for decision-making on projects, examination of laws and legal principles, and governmental development and management of water resources.

3

Assessing Benefits and Costs of Corps Projects

FEDERAL WATER RESOURCES PLANNING AND EVALUATION

Benefit-cost evaluations have been part of Corps of Engineers planning studies since the early twentieth century. As discussed in Chapter 2, the Corps began standardizing its more routine economic procedures in the 1920s, providing estimates of project benefits and costs. Growing budgets related to Flood Control Acts in 1917 and 1928 (the latter in connection with the exceptional Mississippi River floods of 1927) created pressures for greater accountability. In 1927, Congress directed the Corps to study all major U.S. river basins in terms of navigation, hydropower, irrigation, and flood control project prospects. Over the ensuing decade, the Corps produced a mass of documents entitled "308 reports" (named after the U.S. House of Representatives document that mandated the studies), in which the agency relied on some quantification of benefits and costs to impose discipline upon its growing programs.

The Flood Control Act of 1936 mandated formal benefit-cost analysis (BCA) within Corps planning studies. One observer has referred to the act as ". . . one of the heroic efforts of the United States Congress to control its own bad habits" (Porter, 1996). The act states:

> [...] the federal government should improve or participate in the improvement of navigable waters or their tributaries, including watersheds thereof, for flood control purposes if the benefits to whomsoever they accrue are in excess of the estimated costs, and if the lives and social security of people are notherwise adversely affected.

Following passage of the 1936 act, several policy and planning milestones have affected ways in which benefits and costs are calculated within Corps planning studies. Examples of key subsequent federal and Corps water resources planning guidance documents regarding benefit and cost calculations include a 1952 document (Circular A-47) from the

Bureau of the Budget; the 1958 report, *Proposed Practices for Economic Analysis of River Basin Projects*" (known familiarly as "the Green Book"), issued by a subcommittee of the Federal Interagency River Basin Committee; Senate Document 97[1], approved by President Kennedy in May 1962; and the 1973 *Principles and Standards* (*P&S*) and the 1983 *Principles and Guidelines* (*P&G*), both issued by the federal Water Resources Council (WRC, 1973; 1983). Although the methods and criteria for evaluating benefits and costs in Corps studies have changed substantially over the years, BCA has played a central role in prioritizing water project proposals throughout the agency's history and remains the key decision criterion in Corps planning studies. The following section reviews the use of benefit-cost analysis in Corps water resources project planning studies.

BENEFIT-COST ANALYSIS IN WATER PROJECT PLANNING

Benefit-cost analysis is well-established in theory and practice. Water resources planning and research has had important influences on the field of benefit-cost analysis (e.g., Eckstein, 1958; Krutilla and Eckstein, 1958; Maass et al., 1962; Mazmanian and Nienaber, 1979), and there is a broad literature regarding BCA theories and methods for water project evaluation (e.g., Brent, 1998; Garrod and Willis, 1999; Hanley and Spash, 1993; Zerbe and Dively, 1994). Presidential Executive Orders 12,291 (1981) and 12,866 (1993) require benefit-cost analysis of all major executive regulatory proposals that affect human health and the environment. As mentioned, benefit-cost analysis remains the most important criterion in Corps planning studies (USACE, 2000). Benefit-cost analysis is a method for organizing information to support decisions. By imposing discipline and uniformity on the collection, interpretation, and presentation of information, BCA provides a systematic means of assessment, enabling comparisons among projects. It aims to separate acceptable from unacceptable projects (based on objective criteria) to ensure that resources are invested wisely. A secondary role of BCA is to prioritize project alternatives. Examples of decision criteria used in connection with benefit-cost analysis include (1) maximizing the ratio of

---

[1]The document was entitled *Policies, Standards, and Procedures in the Formulation, Evaluation, and Review of Plans for Use and Development of Water and Related Land Resources.*

benefits to costs, (2) maximizing net benefits, and (3) identifying all alternatives for which the benefits exceed the costs. A strict decision rule based solely upon BCA, however, is problematic because the multiple objectives of government agencies like the Corps are not easily reduced to simple criteria. Nevertheless, BCA provides critical information against which public goals can be calibrated and individual projects scrutinized.

Measuring Benefits and Costs

In making choices about how to spend their income and other resources, people make trade-off decisions. Generally speaking, the more something is valued by an individual, the more will a person be willing to give up for it. The concept of *willingness-to-pay* (WTP) as a measure of economic value (benefits) flows from this precept. WTP is a measure of the amount one would be willing to give up in exchange for the good or service being valued. It is the most that one would be willing to forgo, whether or not one actually pays for the good.[2]

Several characteristics of WTP as a measure of value are worth emphasizing.[3] First, it is based entirely on human preferences. There is no innate feature of a good or service, such as its energy content, its biomass, or its vulnerability to extinction, that measures its value, although these characteristics often influence human preferences. Second, WTP is not limited to goods exchanged in markets. To the extent that people care about goods and services that are not traded in markets, such as environmental quality or an endangered species, they will be willing to give up other goods and services to protect them. The valuing of environmental goods is thus consistent with the valuing of goods traded in regular markets, because both applications involve the use of resources to achieve something of value to people. Third, although willingness to pay is typically gauged in monetary units (such as dollars), the concept itself has little to do with money. The key notion is that the value of obtaining something can be measured by how many other goods and services a

[2] To the extent that the price of a good is less than this WTP, consumers enjoy a "consumer surplus"—a value over and above their actual payment.

[3] A related concept of value is willingness to accept (WTA). This is a measure of the minimum amount of other goods and services or money an individual would be willing to accept in exchange for giving up a good or service. Although WTP is used in this discussion for simplicity, the concepts apply equally to WTA.

person is willing to do without. The use of money as the unit of measure is convenient but not essential—it simply allows goods and services to be valued by the same measure. The cost side of BCA involves the full *opportunity cost* of a project (OMB, 2000), which represents what society gives up to obtain project benefits. This includes both direct financial costs and indirect benefits forgone associated with a project. For example, the full opportunity cost of a project might include construction costs, administrative costs, the value of lost environmental benefits, and the economic value of discomfort or inconvenience suffered as a consequence, such as waterway traffic delays during construction.

Benefit-Cost Analysis in Decision Making

In a seminal 1996 essay, 11 prominent economists framed appropriate roles for benefit-cost analyses in environmental and health policy making (Arrow et al., 1996), stating (see also Box 3-1):

> Benefit-cost analysis can play a very important role in legislative and regulatory policy debates on improving the environment, health, and safety. It can help illustrate the tradeoffs that are inherent in public policymaking as well as make those tradeoffs more transparent. It can also help agencies set regulatory priorities.
>
> Benefit-cost analysis should be used to help decisionmakers reach a decision. Contrary to the views of some, benefit-cost analysis is neither necessary nor sufficient for designing sensible public policy. If properly done, it can be very helpful to agencies in the decision-making process.
>
> Benefit-cost analysis should be required for all major regulatory decisions, but agency heads should not be bound by a strict benefit-cost test. Instead, they should be required to consider available benefit-cost analyses and to justify the reasons for their decision in the event that the expected costs of a regulation far exceed the expected benefits. Agencies should be encouraged to use economic analysis to help set regulatory priorities. Economic analyses prepared in support of particularly important decisions should be subjected to peer review both inside and outside government.

BOX 3-1
Principles for Appropriate Use of Benefit-Cost Analysis

Arrow et al. (1996) recommended that all benefits and costs of a project should be identified; those that can reliably be monetized should be, uncertainty in the benefits and costs should be calculated and reported, and distributional consequences (who gains and who loses) should be addressed and reported. By providing this information, not only the final users of BCA will know the best estimates of the net benefits of the project, but they also will have a good understanding of who pays for it and who benefits from it. The principles of benefit-cost analysis they identified are as follows:

- BCA is a useful way of comparing the favorable and unfavorable effects of proposed policies.
- Decision makers should not be precluded from considering the economic costs and benefits of different policies.
- BCA should be required for all major regulatory decisions.
- Although agencies should be required to conduct BCA for major decisions and explain why they have selected action for which reliable evidence indicates that expected benefits are significantly less than expected costs, those agencies should not be bound by strict benefit-cost tests.
- Benefits and costs of proposed policies should be quantified wherever possible. Best estimates should be presented along with a description of uncertainties.
- The more external review regulatory analyses receive, the better are they likely to be.
- A core set of economic assumptions should be used in calculating benefits and costs.
- Although BCA should focus primarily on the overall relation between benefits and costs, a good analysis will also identify important distributional consequences.

SOURCE: Arrow et al. (1996).

Reasons for rejecting a strict benefit-cost test as a decision rule for ranking investment priorities include the following: (1) there may be important equity considerations in the distribution of costs and benefits that are not addressed by maximizing the difference between total benefits and total costs, (2) benefit and cost estimates may contain significant uncertainties, and (3) it may not be possible to use money as a measure of all relevant costs and benefits (e.g., biodiversity, ethical issues).

Consistent with these principles, many economists and planners agree that benefit-cost analysis is an important means for informing public policy decisions, but that it should not be used as a strict decision rule. That is, benefit-cost analysis can be used to identify whether an action increases aggregate well-being, which is its traditional role. Yet benefit-cost analysis could also provide information regarding the distribution of benefits and costs, the robustness of benefit and cost estimates in the face of future uncertainties, and whether nonmonetized factors are important. When BCA is used as a strict decision criterion, however, the richness of these multiple considerations is lost. Benefit-cost analysis is most appropriately employed mainly as a method to inform and support decisions, not as a precise decision rule. Pursuant to the 1936 Flood Control Act, however, the Corps is required to use benefit-cost analysis as a strict decision rule in recommending projects to Congress. This reflects a view that BCA assessment is a specific, strict decision rule, which is inconsistent with principles of "best practice" in modern economics and inconsistent with the principles identified by the Arrow et al. group (1996).

Nonmarket Goods and Services

Many of the goods and services delivered by water resources project are not exchanged in markets. Examples of nonmarket goods and services include some recreational opportunities (e.g., swimming in a lake or river), ecosystem services (e.g. genetic diversity; waste assimilation capacity), and reduced risk of injury or death. Methods exist for including nonmarket goods in benefit-cost analysis (Bateman et al., 2002; Bjornstad and Kahn, 1996; Champ et al., 2003; Freeman, 1994; Haab and McConnell, 2002; Kopp et al., 1997; Randall, 1999; Smith, 1996), but there are methodological challenges associated with consistently valuing them.

The value of environmental goods and services has been the subject of numerous studies.[4] Methods used to obtain quantitative values can be divided into two categories: (1) *revealed preferences* and (2) *stated preferences*. *Revealed preferences* are methods that infer willingness to pay from related market behavior. An example of a revealed preference study is an estimate of the WTP for clean air by comparing real estate prices in more- and less-polluted regions of a city. Similarly, the WTP for more abundant fish populations can be derived from the demand for recreational fishing trips. The unifying theme of revealed preferences is that data on market behavior are used to calculate the value of associated environmental goods.

Stated preferences are methods that elicit WTP directly from individuals through surveys, interviews, or simulated exercises. Begun in the 1960s, stated preference methods have become increasingly popular and today constitute a large literature within the environmental economics field. Much of the appeal of these methods stems from their flexibility. Stated preference methods can be applied to any good or service, at any place or time, including hypothetical goods such as a new recreational opportunity or restoration of a degraded ecosystem. They may also be the only practical approach for some categories of benefits and costs.

Corps guidance recognizes the utility of both revealed and stated preference methods for planning studies (USACE, 2000) and uses them for some categories of benefits. For example, the Corps often uses *unit day values*, estimated through *benefits transfer*, to represent the economic value of environment or natural resources, and the Corps publishes unit day values for use in valuing recreational impacts of its projects.[5] These approaches estimate WTP for project or resource use outcomes using goods or services other than the ones for which the values are initially needed. As an example of the benefits transfer method, assume that a proposed project is being considered that would eliminate fishing in a Louisiana estuary. Rather than conducting a new study of the estuary, an existing stated or revealed preference study on recreational fishing in a Texas estuary could be consulted. Suppose the Texas study found that on average, an angler was willing to pay $25 per day for a fishing outing. If the study is sound and the resulting value is within

---

[4] Braden and Kolstad (1991), Freeman (1993), and Herriges and Kling (1999) provide overviews of methods of nonmarket valuation. Anderson and Kobrin (2002) review the use of benefit cost analyses within the U.S. Environmental Protection Agency's programs.

[5] For example, effective in March 2002, Corps guidance provides for general recreation values ranging from $2.90 to $8.69 per day and specialized recreation values from $11.46 to 34.41 per day (USACE, 2002). A point system for identifying the quality of the recreational experience allows analysts to select values within these ranges.

the range of the Corps' published values for this experience, it might then be "transferred" or applied to the Louisiana project. The number of angling days lost to the Louisiana estuary could then be estimated and multiplied by $25 to yield an estimate of the lost value of the recreational angling. There is an underlying assumption that Texas anglers and Louisiana anglers value estuarine fishing the same way.

Some nonmarket goods and services do not have to be consumed or experienced in order to be valued. For example, some people may value the existence of whales in the sea although they may never see a whale or consume whale-based products. Such "non-use" values can be difficult to accurately measure, as it is difficult to determine how many people will have positive non-use values for a particular good. It has also been argued that such goods might take on symbolic value—valued not so much for themselves but for a broader concern for environmental protection or species conservation that they represent.[6] Nonetheless, there is little debate that non-use values are a legitimate, conceptual component of environmental benefits and should be included in calculations of overall project, regulatory, or programmatic benefits.

Benefit-Cost Analysis in Other Agencies and Organizations

Despite the challenges of valuing environmental goods and services, the Office of Management and Budget (OMB) places enough confidence in established methods to have authorized carefully-constructed valuation studies as important elements of rule-makings (OMB, 2000; 2003; see Appendix C). These guidelines are applicable to all executive-level agencies.[7] The U.S. Environmental Protection Agency (EPA) and the National Oceanic and Atmospheric Administration (NOAA) are mandated to address environmental issues and problems; their experience with valuation is thus relevant to the Corps' ecosystem restoration efforts. In 2000, the EPA issued *Guidelines for Preparing Economic Analyses*, which serves to guide BCA applications for EPA's environmental regulations and programs (similar to the Corps *Planning Guidance Notebook*). The EPA guidelines reflect an awareness of the challenges of measuring and monetizing benefits and costs of environmental

---

[6] This is sometimes called the "warm glow" effect in the environmental economics literature.
[7] Both revealed and stated preference methods have also been used by various federal and state agencies in litigation to recover damages for environmental impacts.

goods and services. Nevertheless, its guidelines state that "to the extent feasible and warranted by their contribution to the results, as many of the effects of a policy as possible should be monetized. This enhances the value of the conclusions to policy makers weighing the many, often disparate consequences of different policy options and alternatives" (EPA, 2000, p. 176).

Within the NOAA, environmental valuation has been an important method for implementing provisions of the Oil Pollution Act that require compensation for environmental damages due to oil spills. In the process of assessing losses associated with natural resources damages, NOAA requires monetary compensation for non-use (often referred to as "passive losses"), a decision that has withstood legal challenge.[8] Nonmarket valuation is also practiced by some international organizations, such as the World Bank, which requires benefit-cost studies of proposed investments (see Box 3-2).

In comparison to other federal environmental and natural resources management agencies (e.g., EPA, NOAA), the Corps has made less use of environmental valuation techniques in its benefit-cost analyses. This may have been appropriate in an earlier era when Corps authorities and U.S. federal laws only tangentially addressed environmental goals. Today, however, ecosystem planning and restoration are major and growing foci of the Corps work program. The Corps' largest restoration projects involve years of planning and evaluation and expenditures of billions of dollars. In this context, current Corps guidance concerning the use of environmental valuation techniques is out of date. Carefully used, these techniques can improve the Corps' planning and evaluation capabilities. The Corps Institute for Water Resources has led agency efforts in supplementing its guidance in this analytical realm (Stakhiv et al., 2003).[9]

[8] *General Electric Co.* v. *U.S. Dep't of Commerce, Nat'l Oceanic & Atmospheric Admin.*, 128 F.3d 767 (D.C. Cir. 1977).

[9] Although it acknowledges the presence of estimation methods suitable for valuing some ecological services, the Corps is guarded in its assessment of the usefulness of these methods:

> . . . considerable technical obstacles, both scientific and economic, stand in the way of comprehensive monetary accounting of restoration project benefits. Scientific obstacles relate to problems in tracing the links between restoration actions and service outcomes underlying all possible routes to human benefits. Economic obstacles relate to methodological limitations for measuring nonmarket benefits of service outcomes that affect the quality of human life in ways that have no close connection to the use of market goods. Together, the scientific and economic obstacles to comprehensive valuation of restoration outputs impede use of a net benefits criterion for justifying restoration projects (Stakhiv et al., 2003).

BOX 3-2
Environment and BCA at the World Bank

The World Bank makes nearly $20 billion in project and program loans annually to nations around the world, approximately ten times the level of Corps of Engineers investment activities. The Bank organizes most of its activities around a "project cycle" that includes both ex ante and ex post evaluations. The Bank's appraisals distinguish "financial" costs and benefits—those that accrue as cash flows to or from the project sponsor—from "economic" costs and benefits that are experienced by all other entities. BCA is accepted within the Bank not only as a systematic way to evaluate proposed investments, but also as a decision criterion, and over the years, Bank economists have made important contributions to the literature on benefit-cost analysis.

The World Bank's guidelines for BCA include the following approach to valuing environmental services: environmental externalities are identified as part of the environmental assessment, quantified where possible, and included in the economic analysis as project costs (e.g., decreased fish catch, or increased illness) or benefits (e.g., reduction in pollution of coastal areas). After monetary values are assigned to costs and benefits, they are entered into the cash flow tables like other costs and benefits (World Bank, 1996). Environmental appraisals are the responsibility of the borrower and are required for all projects expected to have a major environmental impact. The appraisal "predicts and assesses the project's likely positive and negative impacts, in quantitative terms to the extent possible," while the required analysis of project alternatives "quantifies the environmental impacts to the extent possible, and attaches economic values where feasible" (World Bank, 1999). In practice, however, environmental costs are generally not monetized for Bank projects. Recently, some experimental appraisals with environmental monetization have been conducted. For example, the 1999 appraisal of a $3.7 billion oil development and pipeline project in Cameroon and Chad explicitly included a comprehensive net present value of all environmental and social costs—and estimated them at less than $20 million against $1.8 billion in economic benefits. However, although some costs were monetized (e.g., value of lost agricultural and grazing use, loss of edible game harvested in forests), others were merely mentioned (World Bank, 2000).

SOURCES: World Bank (1996, 1999, 2000) and *http://www.essochad.com/Chad/Files/Chad/EAESU1.pdf.*

Implementation of these techniques by the Corps will require approval from the OMB, or revision of the *P&G*, or both.

Nonmarket valuation of environmental goods and services will always be subject to some degree of uncertainty. This should not be a reason to disregard its usefulness in benefit-cost analysis, however. Almost all methods used to evaluate benefits and costs, whether traditional market-based approaches based on engineering studies or international trade studies of projected port usage, are subject to uncertainty. Although nonmarket valuation methods do contain uncertainties and represent substantial analytical challenges, the same could have been said of today's "traditional" methods when they were in early stages of development. These methods were improved over time through repeated applications and refinements. The growing importance of environmental evaluations and projects within the Corps suggests that the relevance and potential usefulness of nonmarket valuation methods will continue to grow. The Corps should thus continue its efforts in improving the scope, reliability, and applicability of valuation methods (an ongoing National Research Council study on the valuation of goods and services within aquatic ecosystems, cosponsored by the Corps and due for publication in 2004, should be useful to the agency). Given the extensive use of benefit-cost analysis within the federal government, the Corps would also be well-served by cooperating with other federal resource management agencies, notably EPA, NOAA, and the Departments of Agriculture and Interior, to coordinate and support research to improve BCA methods and applications. The Office of Management and Budget should support such cooperative efforts.

STATE OF CORPS PRACTICE: CURRENT ECONOMICS APPROACHES

The Corps *Planning Guidance Notebook* (USACE, 2000; see Box 3-3) establishes standards according to which all projects should be scrutinized. These principles of analysis are accepted by many economists and policy analysts, although they do contain some unusual features.

Efficiency

The economic notion of *efficiency* refers to the degree to which the marginal costs of an action or policy equal its marginal benefits. Measures that better match these increases in costs and benefits thus increase economic "efficiency." Principles c, d, and e of the *Planning Guidance Notebook* (Box 3-3) emphasize the use of benefit-cost, optimization, and incremental analyses in attempting to gauge economic efficiency. All three types of analysis are means of determining whether a project's benefits will be greater than its costs and what specific project components maximize net benefits (the difference between total benefits and total costs). These principles state that where benefits are difficult to measure, the costs of attaining a specific outcome should be minimized.

The Corps' planning guidelines emphasize efficiency. Not only must a project produce benefits that exceed costs, but each incremental component of a project must also produce benefits in excess of costs. Plans that produce higher net benefits are preferred. However, "efficiency" is not the Corps' only guiding principle. The *P&G* and *PGN* each state that project plans should also be evaluated according to their completeness, effectiveness, and acceptability. Unfortunately, the guidance documents are largely silent on procedures for documenting project performance with respect to these other principles.

A Mississippi River levee project in St. Paul, Minnesota, illustrates this issue. The project aimed primarily to protect floodplain structures and inhabitants by raising levees, clearly part of the Corps' mission to mitigate flood damages. Beyond simply raising a levee, the design ultimately included decorative stone facades, walking and bicycle paths, public gathering areas, and decorative lighting and ironwork (USACE, 1981). It could be argued that such decorative and recreational refinements satisfy the completeness and acceptability criteria and, as such, are legitimate objectives—they rendered the project more complete in the sense of contributing to broad public values and they improved its acceptability to local sponsors. From the standpoint of economic efficiency, adding refinements to the project may have been the least expensive way of achieving the city's development objectives. Recreational aspects, however, were not included in the original authorization and could not be included in calculating the project's national economic development (NED) benefits. Recreational and aesthetic considerations were thus part of a locally preferred alternative (not the NED alternative), the costs of which were borne by the local sponsors.

The *Planning Guidance Notebook* does not help Corps planners reconcile the completeness, effectiveness, and acceptability criteria with the prevailing criterion of efficiency. Corps planners would benefit from more guidance on this issue. The intent of this guidance should not be to

BOX 3-3
Corps Principles of Analysis

a. **Systems analysis**—Consider broad system aspects of problems and solutions.

b. **With- and without-project analysis**—Evaluate future conditions with and without the project; do not compare the future with the project to the past without it.

c. **Benefit-cost analysis and cost-effectiveness analysis**—Determine all project costs and benefits, monetized or not; if benefits are not easily quantified, determine the costs of alternative ways of achieving a specific objective.

d. **Net benefits (optimization)**—Favor the project for which the benefits are greatest relative to the costs.

e. **Incremental analysis**—Evaluate separable elements of a project individually and cumulatively to ensure that each element adds more than it costs.

f. **Trade-off analysis**—Identify trade-offs between multiple benefits.

g. **Risk and uncertainty**—Recognize and quantify the variability of project outcomes, including trade-offs between risk and cost.

h. **Planning area**—Analyze impacts in the geographic region identified in the authorizing document plus the areas that would be affected by alternative plans.

i. **Prices**—Use the general level of prices prevailing during or just before the planning period for the entire project period. Discount future benefits and costs and compare them at a common point in time using a discount rate published by Corps Headquarters.

j. **Period of analysis**—Use the same period of analysis for each alternative plan—up to 50 years or, for multiple purpose reservoir projects, up to 100 years.

k. **National economic development costs**—Evaluate costs according to the "opportunity cost" principle. Include the direct costs of implementation, the economic costs of resources for which no financial outlays are made, and the costs incurred outside of the project to take advantage project outputs.

l. **Environmental and social impact assessment**—Consider a full range of social and economic effects as required in applicable federal laws such as the National Environmental Policy Act, Clean Water Act, Rivers and Harbors Acts, and Flood Control Act.

m. **Significant resources and significant effects**—Focus on the significant resources and effects as identified, in laws, plans, and policies of government and private groups; in scientific or technical studies, or by a segment of the general public.

n. **Regulatory considerations**—Incorporate applicable provisions of Department of the Army regulatory programs requiring permitting of activities.

o. **Project implementation timing**—The schedule of implementation is a variable in project planning; consider the effects of timing on annual benefits and costs.

p. **Hazardous, toxic, and radioactive wastes**—Projects may not include cleanups regulated by the Comprehensive Environmental Response, Compensation, and Liability Actor the Resource Conservation and Recovery Act or compliance actions required by other laws apart from the proposed project.

q. **Brownfields**—Projects may include the cleanup of small contamination problems that do not reach CERCLA criteria.

r. **Congressional adds**—These provisions apply to all congressionally added studies unless specifically precluded in the budget process.

SOURCE: Paraphrased and summarized from USACE (2000, pp. 2-8 to 2-14).

diminish the importance of efficiency, but to suggest ways in which completeness, effectiveness, and acceptability complement the efficiency criterion. In the St. Paul project, for example, the improved appearance of the project could be expected to increase recreational activities and enhance prospects for complementary development efforts.

The Planning Region and Allocating Costs and Benefits

Principle h of the *PGN*'s principles of analysis—planning area—indicates that Congress or the OMB is to define the region of analysis through the "authorizing document" for a specific project. By comparison, in the 1936 Flood Control Act, benefits were to be considered "to whomsoever they accrue," clearly indicating that the consideration of to whom or where benefits accrued was not to be the basis of a decision. Under principle c, a subsection contains language to this effect: "benefits are counted wherever they accrue (even outside the study area. . .)" (USACE, 2000, Sec. 2-4c (4), p. 2-8). Similarly, provisions for ecosystem restoration in the *PGN* call for changes to be "measured in the planning area and in the rest of the Nation" (Sec. 2-2b, p. 2-2). This presents mixed signals about the importance of the "planning region" relative to an accounting of all benefits and costs to whomsoever they accrue.

To be consistent with the benefit-cost principles described in Box 3-1, it would be appropriate to include all significant economic effects in planning studies and to address legislative and budgetary considerations separately. Including a discussion of who receives the benefits and who incurs the costs would provide better information about the distribution of project benefits and costs. The *Planning Guidance Notebook* should be revised to incorporate these types of more explicit provisions about the spatial and distributional dimensions of costs and benefits.

CORPS MISSIONS AND METHODS

Although the *PGN*'s principles of analysis are not very controversial, their application to individual projects requires interpretation. Each Corps mission area presents distinct analytical challenges, and the *PGN* devotes a 40-page chapter and a 270-page appendix to the nuances of these individual missions. The documents provide a template for planning studies and remind practitioners of best practices. At the same time, it takes planners years to understand how the various guidelines fit together, and ongoing education is a vital part of improving the Corps planning process. The learning curve is especially steep for local sponsors. They must participate in and share the costs of planning studies, but they often do not understand the principles that guide Corps planning and the specific rules that limit the flexibility in some areas. It is typically easier to follow established procedures than to exercise independent judgment, but some degree of judgment is required and often desirable in many aspects of water resources planning; each water project and planning study is different and will not conform to all general planning criteria. This gives experienced, creative analysts an opportunity to design place-specific and innovative alternatives. It also points to the need for the Corps to retain competent staff with knowledge of and education in the engineering, economics, and environmental dimensions of water resources management. The Corps recognizes this and has initiated efforts to this end (Box 3-4).

Navigation

The evaluation framework for navigation projects is underpinned by the clear statement that the "conceptual basis for benefits is willingness-to-pay for each increment of output from a plan" (USACE, 2000, p. 3-4). This is consistent with best practices of benefit-cost analysis. However, in the next paragraph, the guidance notes that "the base economic benefit of a navigation project is the reduction in the value of resources required to transport commodities." The lowered cost of transport is assumed to be the dominant source of social willingness to pay for improved navigation services. Although this may capture many of the benefits, it omits others. For example, the *PGN* does not allow the inclusion of reduced highway fatalities in calculating navigation benefits.

Flood Damage Reduction

With regard to flood damage reduction projects, the *PGN* notes that benefits from these projects derive from three categories: (1) inundation reduction, (2) intensification benefits (net income from affected "with-project" land uses, when floodplain use is the same), and (3) location benefits (derived from activities added to the with-project floodplain). Inundation is the most important of the three, according to the *PGN*:

BOX 3-4
Ensuring the Quality of the Corps Planning Work Force

In July 2000, faced with concerns that its plan formulation capability was diminishing as experienced planners retired or left planning, the Corps chartered a task force of Corps Headquarters, division, and district personnel to develop recommendations regarding the training and development of Corps planning staff. The task force noted that " . . . at the heart of that work force is the planner, a person who is well versed in the problem solving arena and is capable of creating solutions to water resources problems" and that " . . . planning expertise is [currently] vested in an alarmingly small number of people within Civil Works." The task force recommended several actions to "hire-train-retain" planners, and proposed an education program that would require new planners to participate in a series of workshops and offer them opportunities to enroll in other seminars, workshops, and courses.

As a result of the task force recommendations, the Corps initiated a "core curriculum" of basic planning workshops, established a six-month planning associates program in advanced planning skills for mid-career planners, and is embarking on a cooperative program with several universities—Arizona, Florida, Harvard, Johns Hopkins, Southern Illinois, and Washington State—to offer interdisciplinary water resources planning master's degrees to Corps employees. Under the program (at Corps expense), Corps staff attend a participating university for one semester and take a series of required courses. They complete the remainder of courses using locally available courses and distance learning courses. The Corps intends to enroll an average of one planner from each district each year in the program (*http://www.usace.army.mil/ mastersdegree/index.htm*).

These efforts are commendable. However, because most Corps activities are project funded, it is difficult for some districts to support the above programs. The Corps should seek specific authorization and appropriations from the Congress to support such activities in order to ensure their continuity and viability. A well-educated work force is but one part of a sound planning base. Other issues such as salaries, working conditions, hiring procedures, and administrative procedures also play a role in retention and are being considered by the Corps. Progress in these areas, however, will require resources and attention from beyond the Corps. The administration and congress will have to support efforts on these fronts to enhance the Corps' ability to attract and retain competent and qualified professional staff.

"Benefits from plans for reducing flood hazards accrue primarily through the reduction in actual or potential damages to affected land uses" (USACE, 2000, Sec. 3-3c (1), p. 3-14). These guidelines, however, capture only a portion of the potential costs and benefits associated with flood damage reduction projects. One important category relates to human lives and well-being. There are risks to human life associated with even the most reliable flood control structures that should be part of comprehensive benefit-cost analysis. For large flood control projects, these benefits can be substantial, yet they are difficult (if not impossible) to measure according to procedures prescribed within the *P&G* (given the magnitude of the payments to the relatives of those lost in the 9/11 terrorist attacks, for example, benefits for flood damage reduction developed using similar assumptions and values would clearly increase the monetary value of projects). Another category relates to environmental variables such as habitat that are affected by flood damage reduction projects. Structural and nonstructural flood control projects will affect these categories in different ways, as nonstructural flood damage reduction projects often seek to simultaneously reduce flood damages and enhance environmental benefits. If these environmental benefits are not considered, however, there may be an analytical disincentive against these types of project.

Measuring Benefits of Nonstructural Measures

The *PGN* requires the Corps to consider nonstructural alternatives to achieve project goals for flood damage reduction projects. Nonstructural measures for reducing flood damages generally do not modify flows of flood waters, and include measures such as relocating structures out of the floodplain, the floodproofing or elevating of buildings, and warning systems. Calculating the costs and benefits associated with nonstructural measures poses analytical challenges, partly because such measures often involve "disinvestments" rather than investments. The relocation of structures from a floodplain, for example, typically reduces the market value of the associated land because of the removal of capital investments. In conventional economics terms (such as those embedded in property taxation formulas), the land is considered less productive without those buildings. In conventional benefit-cost terms, the relocation of a structure would be warranted only if it makes way for a more economically productive use of the site. Nonstructural alternatives may also entail benefits that extend beyond a project's locale, which poses further analytical complications. For example, relocating structures from a floodplain may reduce the value of a parcel of land, but removal may also increase the value of other properties if floodwaters are allowed to flow into the vacated area, thereby reducing downstream peak flows and flood damages. The increased wildlife habitat associated with the removal of structures could also increase the value of adjacent properties. There are also potential gains in nonmarket values, such as biodiversity.

One type of nonstructural flood management project entails relocation of structures out of the floodplain. According to Corps policy on floodplain relocations (USACE, 2000, Appendix E, Sec. III E-17b (1), p. E-84), relocations must be justified largely by the offsite and nonmarket benefits (Shabman et al., 1997).[10] In addition, the *Principles and Guidelines* lists the benefits to be claimed for "evacuation" projects—the value of flood damages reduced in these projects are not claimable. These policies presume that the costs of flood damages are capitalized into the value of the structures and that the cost of purchasing floodplain structures thus already incorporates the benefits of relocation. So-called double counting of benefits is a reasonable concern and should be avoided, but the Corps policy and the *P&G* presume that real estate markets work perfectly to reflect the risk of possible flood-related damages. Empirical studies, however, suggest that the risks of floodplain occupation are not fully reflected in real estate values. In an investigation of the effects of flood hazards on real estate prices, it was concluded that, "A general discount for floodplain *location* . . . does not exist (Chao et al., 1997; see also Chivers and Flores, 2002[11]). A previous National Research Council committee also investigated this issue in some detail, concluding that ". . . the benefits of flood damages avoided (should) be included in the benefit-cost analysis of all flood damage reduction projects—including all nonstructural projects . . ." (NRC, 1999a).

An example of a prominent nonstructural Corps flood management project was the permanent evacuation of 130 residential structures and two businesses (as well as some floodproofing of remaining structures) at Prairie du Chien, Wisconsin, in 1978. Prairie du Chien is located on the

---

[10] The benefits are the differences in land rents with versus without the project, plus any off-site benefits that might result and any savings in administration cost for polices in the national flood insurance program that are no longer in effect. The land rents will often be reduced, in which case the use of nonstructural measures will depend on the off-site and insurance benefits.

[11] The authors recommend that sellers of properties that participate in the National Flood Insurance Program be required to include a flood elevation certificate and a flood insurance cost quotation at the time of listing.

left bank of the Mississippi River just north of the mouth of the Wisconsin River. Floods at Prairie du Chien had been frequent and devastating. The record Upper Mississippi flood of 1965, for example, caused damages exceeding $1.9 million (Moore and Moore, 1989). Following completion of the project, a benefit-cost reanalysis was conducted, which concluded that "the traditional benefit/cost procedures were inadequate when measuring nonstructural projects such as Prairie du Chien" (Miller et al., date and Klemme, date as cited in Moore and Moore, 1989). Frequent deployment of federal and state disaster relief would also lead one to believe that floodplain real estate values may not be discounted by the full economic cost of potential floods (see NWF, 1998, for detailed discussion of repetitive losses from flood damages). Moreover, a large body of research (e.g., Holoway and Burby, 1990; White, 1945) has concluded that people systematically underestimate the risks of flooding.

As far back as 1970, the Corps had formed an ad hoc group of staff members to meet periodically to discuss nonstructural approaches to flood damage reduction (in 1985 this group was formally chartered as the National Floodproofing Committee). Following the 1993 Mississippi River floods, the National Floodproofing Committee served as a catalyst for efforts to consider broader applications of nonstructural approaches. The group was rechartered in 2003 as the National Nonstructural/Floodproofing Committee. The committee has been active in preparing floodproofing guides for local governments and private citizens and in promoting nonstructural alternatives for managing flood risks. As the Federal Emergency Management Agency (FEMA) moved forward with relocations following the 1993 Mississippi and Missouri river floods, the Corps expanded the use of relocations in planning flood projects (USACE, 1999b). One example was the relocation of 300 families from homes and apartments along the Red River in Grand Forks, North Dakota, and East Grand Forks, Minnesota (USACE, 2001c). These relocations moved residents out of harm's way and created a wider path for floodwaters to dissipate and move downstream in the Red River. The Corps views relocation as a viable approach to managing flood risks, and other nonstructural measures are in planning or construction at more than 50 locations (USACE, 1999b; 2001c).

Nonstructural measures pose analytical challenges because they are frequently associated with changes in policy. Many floodplain structures were built in an era when governments subsidized flood protection or compensated private landowners for most losses. Some of these subsidies promoted efficiency (i.e., the increase in the productive value of the land exceeded the costs, both private and public). Others were motivated by distributional or political considerations. The result of the latter policies, in particular, was to inflate the relative value of floodplain land in developed uses and thereby induce more construction. In some cases, the nation would be better served not by constructing additional structures in floodplains, but rather by promoting development in areas less prone to floods. However, when projects are evaluated incrementally (as they tend to be under current evaluation procedures) rather than in a more comprehensive sense, and when past projects have inflated property values, it is difficulty to deny the next project proposed to protect those high property values.

The reasons the Corps has not had a more vigorous nonstructural flood damage reduction program are complex, as noted in the NRC (1999a) report:

> There appears to be a large and increasing demand for Corps-sponsored nonstructural flood damage projects . . . .There is an apparent mismatch, however, between this perceived demand and the federal response. The reasons for a relative lack of Corps-sponsored nonstructural projects are not clear. This may be the result of skewed benefit calculation procedures; it may be imbedded in an institutional bias against nonstructural projects; it may be that Congress and the Office of Management and Budget do not see a federal interest in local nonstructural projects.

Another report also commented on a limited emphasis on nonstructural approaches to managing flood risks:

> Structures had been used for generations and their costs and benefits were well understood. Their physical presence instilled a source of security. Their effects were permanent and, with periodic monitoring, predictable throughout the life of a project. Ultimately, structures removed individual decisions from the political agenda and validated generalizations about land use.
>
> By contrast, nonstructural measures kept people away from the water, rather than water away from the people. They employed unfamiliar and nontraditional activities like zoning and flood preparedness, which require

> personal involvement, and they called for individual sacrifice, such as paying for flood insurance. (Moore and Moore, 1989).

Other Federal Agencies

Many federal agencies other than the Corps play important roles in reducing pre- and post-event impacts of floods. Each agency has its own specific mission, some part of which involves reducing damages from floods. Some of these agencies—namely, the Bureau of Reclamation, the Natural Resources Conservation Service, and the Tennessee Valley Authority—follow the *Principles and Guidelines*. Other agencies with important roles in flood impacts—such as the Department of Housing and Urban Development and the Small Business Administration—follow procedural guidelines other than the *P&G* because their programs and projects are generally smaller and more localized. Another agency that operates beyond the jurisdiction of the *P&G* is FEMA, which promotes preparation for and mitigation of flood damages. Among its various flood-related activities (which include oversight of the National Flood Insurance Program), FEMA provides guidance on benefit-cost calculations to help determine the cost-effectiveness of flood damage mitigation projects (*http://www.fema.gov/fima/hmgp/riverine_a.shtm*; accessed January 24, 2004). This site lists specific benefit categories allowed for hazard mitigation projects: building damages, contents damages, displacement costs, business income losses, rental income losses, and lost public or nonprofit services. FEMA is not obligated to follow a specific, federally-defined evaluation process. An example of FEMA's benefit-cost guidance for proposed projects is as follows (FEMA, 2003, pp. 1-2; emphases in original):

> the focus of hazard mitigation projects is on strengthening, elevating, relocating or otherwise improving buildings, infrastructure or other facilities to enhance their ability to withstand the damaging impacts of **future** disasters . . . .The **benefits** considered are **avoided** future damages and losses which are expected to accrue as a result of the mitigation project. The **costs** considered are those necessary to implement the specific mitigation project under evaluation . . .
>
> The benefits considered in the **Benefit-Cost (BC) Program** include: avoided damages to the building and

contents, avoided displacement costs, avoided rental and business income losses, and avoided loss of public/nonprofit services

These guidelines provide no indication that FEMA excludes the benefits of avoided damages *for nonstructural alternatives* (such as relocations), as the Corps does. There thus appears to be notable differences between Corps and FEMA policies toward nonstructural solutions. FEMA's focus is on a generalized benefit-cost analysis that supports citizen well-being, as reflected in the reduction of the number of people at risk, while the Corps follows the procedures defined in the *P&G*.

Summary

Current benefit calculation procedures for Corps flood damage reduction projects include only a subset of potential benefits. This is inconsistent with the spirit of benefit-cost analysis, however. Benefit-cost analysis should aim to count all benefits and costs of a proposed action for inclusion in the final analysis, not just a subset of preferential categories or benefits that may be consistent with conventional methods for their calculation. If a flood damage reduction project affects human lives and the environment, these respective costs and benefits should be noted in the analysis. Past studies have shown that the benefit calculation procedures used by the Corps in its flood damage reduction projects—especially nonstructural projects—do not consistently capture the full range of benefits. These limitations may be discouraging the agency from greater involvement in these types of projects. Several other federal agencies have responsibilities for addressing flood-related damages. Some of these agencies are guided by the *Principles and Guidelines*; others are not. The resulting mixture of approaches to calculating confuses the public and likely confuses the agencies, as well. The issue of ensuring consistency of methods and applications among federal agencies in this realm merits careful consideration by the administration and Congress.

Ecological Restoration

The late twentieth century saw shifts in the treatment of environmental benefits in Corps planning studies. The 1972 *Principles and*

Standards, for example, established four accounts in Corps planning studies, one of which was "environmental quality," which was treated as a coequal objective to national economic development (NED; see Chapter 1). The 1983 *Principles and Guidelines* marked a policy shift because it required that only one account—national economic development—be established in Corps planning studies. The principles in the *P&G* relegated environmental considerations to a constraint, stating that the federal objective in water and land resources planning was to "contribute to national economic development consistent with protecting the Nation's environment, pursuant to national environmental statutes, applicable executive orders, and other Federal planning requirements" (WRC, 1983). This policy change meant that the Corps, where it believed it was feasible, had to identify, measure, and monetize the environmental effects of its projects if they were to be included within the NED account. However, given, the reduced importance of the environmental account vis-à-vis the NED account, the Corps was under little pressure from the administration to develop techniques for monetization of environmental goods and services.

The Corps' ecological restoration mission marked another policy shift with regard to the environment and increased the Corps' attention to environmental benefits, rather than only environmental costs. During the 1990s, to help incorporate environmental benefits into the Corps planning framework, the Corps and its Institute for Water Resources sponsored and conducted several studies regarding the accounting of environmental benefits (e.g., Apogee Research Inc., 1996; IWR, 1996, 1997; Martin and Stakhiv, 1999; Shabman, 1993). As previously discussed, incorporating environmental benefits into the traditional Corps planning framework, which is geared toward valuation of the traditional benefits of Corps water projects, represents an analytical challenge to the agency.

The Corps 2000 *Planning Guidance Notebook* presented another important shift in environmental benefits analysis. In the *PGN*, a new planning account—national ecosystem restoration—was established: "For ecosystem restoration projects, a plan that reasonably maximizes ecosystem restoration benefits compared to costs, consistent with the Federal objective, shall be selected. The selected plan must be shown to be cost-effective and justified to achieve the desired level of output. This plan shall be identified as the National Ecosystem Restoration (NER) Plan" (USACE, 2000). Furthermore, and in a marked departure from the *P&G*, this NER account does not require projects to exhibit NED benefits.

Defining and Valuing Benefits

Definitions Environmental restoration is an important concept to the Corps, to ecological scientists, and to many interest groups and citizens. The concept gained in importance and prominence during the 1990s, and its definition and applications evolved during that decade (see Box 3-5).

The field of ecosystem restoration is supported by an increasingly rich theoretical base, but there has been limited experience and limited evaluation of results to date. Establishing an operational definition of this broad concept that is consistent with and complementary to existing

BOX 3-5
Evolution of the term "Restoration"

Restoration: returning an ecosystem to a close approximation of its condition prior to disturbance (NRC, 1992).

Rehabilitation: modifying selected sections of riverine systems to a predetermined structure and function (Gore and Shields, 1995).

Naturalization: shifting some characteristics of the regulated system closer to a natural pattern while maintaining or enhancing economic and social uses of the system (Rhoads and Herricks, 1996).

Normalization: the standard established from what is possible in a natural-cultural context as opposed to pristine conditions, which are difficult, if not impossible, to define or achieve (Stanford et al., 1996).

Restoration: returning a site to a condition similar to the one that existed before it was altered, along with its predisturbance functions and related physical, chemical, and biological characteristics. The goal is to establish a site that is self-regulating and integrated within its landscape, rather than to reestablish an aboriginal condition that can be impossible to define and/or restore (Middleton, 1999).

Restoration: the process of assisting the "recovery" of an ecosystem that has been degraded, damaged, or destroyed . . . interact with contiguous ecosystems including cultural (SER, 2002).

guidance and programs is a challenge for an action-oriented agency such as the Corps. Although there is not an agency-wide, standard definition of restoration, definitions that are used are generally consistent with contemporary literature on the topic. For example, according to internal Corps guidance:

> Civil works ecosystem restoration initiatives attempt to accomplish a return of natural areas or ecosystems to a close approximation of their condition prior to disturbance, or to a less degraded, more natural conditions. In some instances a return to pre-disturbance conditions may not be feasible. However, partial restoration may be possible, with significant and valuable improvement made to degraded ecological resources. The needs for improving or re-establishing both the structural components and the functions of the natural area should be examined. The goal is to partially or fully restore the attributes of naturalistic, functioning and self-regulating systems (USACE, 1999b).

Although consistent with scientific literature on the topic, such definitions do not provide specific guidance for Corps planners. As a result, the Corps may find it difficult to define its limits and strengths within the broad realm of ecological restoration. For example, ecological restoration could include changes in dam operations, reintroduction of native species, changes in water quality, changes in land use, and the creation of new wetlands. Many of these issues were addressed in a National Research Council report (NRC, 2002a) that examined the scientific, as well as organizational, challenges facing the Corps in its efforts to balance a variety of uses and mandates on the Missouri River, which include the protection of endangered species (Box 3-6). That report also pointed out the need for Congress to review and address the multitude of laws and guidance that it has issued pertaining to management of the Missouri River dam and reservoir system.

Valuation Techniques As subjects for economic analysis, ecological restoration projects share some similarities with other Corps projects, as they entail an initial investment, various subsequent expenditures, and a stream of benefits valued by humans. Some characteristics common to most ecological restorations, however, should be stressed lest they be

BOX 3-6
The Corps and Missouri River Management

The Corps of Engineers constructed and operates six mainstem dams on the Missouri River. Those dams are the heart of North America's largest reservoir storage system and exert considerable control on the river's hydrology. Constructed to enhance Missouri River navigation and reduce flood damage, the dams have also altered the dynamics of the river's hydrologic and geomorphic systems, which has contributed to the declining condition of the river-floodplain ecosystem.

The Corps' main document guiding system operations is the *Missouri River Mainstem System Reservoir System Regulation Manual*, or the "Master Manual" (USACE, 1979). The first Master Manual was issued in 1960. In response to a severe drought across much of the basin in the late 1980s, the Corps initiated a process of revising the Master Manual, a process that was not yet completed when this report went to press. Operations of the system represent a delicate and controversial balancing act for the Corps, because originally authorized purposes of the system and subsequent environmental laws may pull the agency in different directions. As General David Fastabend, Commanding General of the Corps' Northwestern Division explained to the U.S. House of Representatives, "Our guidance is sometimes contradictory and the resolution of those contradictions is extremely problematic" (Fastabend, 2002).

In 1999, the U.S. EPA and the Corps requested the National Research Council to convene a committee to provide advice on Missouri River ecosystem science. There have been some efforts aimed at enhancing the Missouri's environmental benefits, but as the NRC committee noted in its 2002 report, "Degradation of the natural Missouri River ecosystem is clear and is continuing" (NRC, 2002). That report also pointed to the key factors that must be addressed in order to effect restoration: "Degradation of the Missouri River ecosystem will continue unless some portion of the hydrologic and geomorphic processes that sustained the pre-regulation Missouri River and floodplain ecosystem are restored—including flow pulses that emulate the natural hydrograph, and cut-and-fill alluviation associated with river meandering" (NRC, 2002a).

neglected in the analysis. First, ecological restoration tends to involve multiple benefits from a single complex natural entity. For example, a restored wetland can simultaneously reduce downstream flooding, trap sediment, purify water, produce hunting and fishing opportunities, provide attractive scenery, and support endangered species. Most, although not all, of these benefits are nonmarket benefits, so their value must be calculated by revealed or stated preferences, rather than by market prices. Also, some benefits may be consumed by people as existence values rather than use values, and geographically, it may be difficult to delineate precisely where each benefit is enjoyed.

Second, the benefits are produced by natural systems, and the complexities of these systems confound precise forecasts. According to guidelines issued by the Society for Ecological Restoration (SER, 2000, p. 7), "no two intact ecosystems are ever identical . . . (and) no restored ecosystem at a project site can ever be identical to any single reference" . (Box 3-6 lists attributes of restored ecosystems, according to the SER). These values are thus distinct, although from an economic point of view, there may be reasonable substitutes for specific habitats when it comes to producing specific benefit flows of value to people and societies. Moreover, ecosystems change and develop over time and may still function despite natural or human perturbations. This makes them different from most engineering investments, in which there is a mechanistic relation between a fixed piece of capital and a stream of benefits.

Third, ecologists view "restoration" not as a particular state, but rather as "attempts to return an ecosystem to its historic trajectory" (SER, 2002). The idea of trajectory implies that an ecosystem may require years to (re)gain its ability to produce a full array of benefits. Ecosystem restoration could thus be disadvantaged in benefit discounting, compared to structural approaches that generate ecological benefits more quickly, or even to simplified, constructed ecosystems such as wetlands populated with nonnative, rapidly growing species. If a natural ecosystem provides distinctive services that are more highly valued than an artificial environment, a proper accounting of service flows and their values may outweigh timing issues (Box 3-5). In giving fair analytic treatment to a restored system, it is essential to consider the range of trade-offs between benefits. For example, nonnative species may quickly control and help manage runoff in wetlands, but as a result, the wetland may never regain its previous state of wildlife values.

Discounting Most Corps projects have design lives of many decades. To permit comparison of benefits and costs accruing at different times, a means of normalizing benefits over time is required. Normalizing benefit and cost streams also permits comparisons between projects with different life cycles. Benefit-cost analysis uses discounting procedures to normalize financial outcomes over time. The logic is that resources are productive and time is valuable. In the marketplace, the interest rate signals the marginal productivity of financial resources and the marginal willingness of individuals to defer consumption to some future date. Implicitly, benefits obtained sooner are worth more than those same benefits realized later, a difference reflected by the amount of interest that can be earned in the interim. Future benefits and costs can be normalized to present benefits and costs by discounting at the rate of interest.

Although discounting is well established in its treatment of financial resources,[12] the Corps is reluctant to apply discounting to physical resources yielding streams of benefits over time, such as habitat or the members of an endangered species. The *Planning Guidance Notebook* (USACE, 2000, Sec. E-36c (1), p. E-154) states that "ecosystem restoration outputs are not discounted, but should be computed on an average annual basis, taking into consideration that the outputs achieved are likely to vary over time."[13] Temporal considerations are relevant to all productive resources, however, whether the services yielded can be monetized by standard means (e.g., flood damage reductions) or not (e.g., enjoyment of a scenic vista). It does not seem wise to reject the potential value for discounting out of hand simply because the units being discounted are physical rather than financial. The Corps may wish to reevaluate its cost effectiveness and incremental cost analysis to help address these issues.

---

[12] The issues surrounding discounting are important for the Corps. There is an extensive literature on the theory of discounting (e.g., Lind et al., 1982). At the same time, the Corps must follow the guidance issued by the Office of Management and Budget.

[13] Stakhiv et al. (2003, p. 92) also address the challenge of discounting.

OTHER ISSUES

The Challenges of Forecasting: "With" and "Without-Project" Forecasts

Perhaps the single most difficult methodological challenge facing project analysts is forecasting future conditions. The value of many Corps proposals often hinges on the size and scope of future economic activity. Deepening a harbor or navigation channel is often worthwhile only if shipping costs are reduced or if the amount of commerce increases enough to more than offset the costs associated with dredging. However, future cost savings and commodity flows will depend not only deepening the harbor, but also on future competition from other nations and ports, technological or regulatory changes affecting competing modes of transport, energy costs, economic growth rates, and other factors. Modest changes in some of these factors can have marked effects on the economic development benefits of a navigation project. Similarly, the forecasting of future environmental conditions is confounded by the roles of extreme events, human influences, and uncertainty regarding long-term environmental changes and variability.

In July 2002, the Corps issued a new policy precluding the use of "locally developed planning models which cannot withstand national level scrutiny" (Griffin, 2002). According to the policy declaration, The Corps Institute for Water Resources will validate Corps economic forecasting models through an independent expert review process. Although this is a step in the right direction, it falls short of the practices by other federal agencies that have instituted more routine review processes that entail a greater degree of independence. The U.S. Department of Agriculture (USDA), for example, supports the Food and Agricultural Policy Research Institute at the University of Missouri, and the National Science Foundation (NSF) supports the National Center for Atmospheric Research in Boulder, Colorado. Both of these institutions develop and manage complex models that are used extensively by their respective sponsoring agency, as well as by other agencies. The Corps Engineering Research and Development Center (ERDC) and Institute for Water Resources, for example, have produced many highly regarded engineering and scientific models and reports. Corps models and reports, however, have not regularly been subjected to fully independent reviews by experts from outside the agency. Such independent reviews are becoming an increasingly important component of the credibility of federal agencies and other bodies.

External Expertise

The relationships that some other federal agencies have with external experts ensure that relevant modeling capabilities are available to the agencies, that expertise accumulates, and that research and studies are subjected to the scrutiny of external experts (see also the report from the 216 study panel on peer review methods; NRC, 2002). Analyses in Corps planning studies today are highly sophisticated, suggesting that the Corps would benefit from the advice of experts outside the agency. Moreover, the Corps is often subjected to strong national- and local-level pressures in its more controversial and complex studies. Given the complexities and pressures that surround the Corps' larger and more costly planning studies, there is the need to involve independent experts to ensure that its plans and projects are technically sound, as well as to ensure a higher degree of credibility.

Other federal agencies also routinely engage outside experts to develop and improve analytical tools. For example, the long-range timber supply and demand models used by the U.S. Forest Service were developed with university economists and have been the subject of numerous peer-reviewed articles. The Corps has engaged independent experts at its research centers at the Institute for Water Resources and the Engineering Research and Development Center; however, the use of independent expertise has been less common in the Corps of Engineers than in some other agencies. Experts can be used to conduct many aspects of the planning studies or at least to serve on advisory bodies charged with ensuring that defensible methods and assumptions are used. They can also be called upon to recommend planning approaches, techniques, and tools and to assist operating staff in their use. Routine use of external experts and carefully-reviewed planning methods will better enable the Corps to defend its studies. The agency is today keen on enhancing the credibility of its planning studies, and in many cases, there is no substitute for the credibility that attends independent, expert review. Increased participation of external experts may also help the Corps contend with a long-term decline in the size of its planning staff that is likely to be compounded by impending retirements. Finally, systematic peer review of Corps planning studies would help ensure that the methods used represent best practices, that assumptions are reasonable and justifiable, and that plausible options for achieving national goals are not being overlooked.

Scaling Analysis to Project Importance and Cost

A reasonable principle of policy or project analysis is that the intensity and cost of a study should be commensurate with the scale and cost of the proposal. Apart from the general exhortation in the introduction to the *PGN* to "be guided by common sense in applying the policies and procedures contained herein" (USACE, 2000, p. 2-1), this principle is explicitly stated only in Appendix F dealing with the Continuing Authorities Program ((USACE, 2000, Sec. F-3b):

> Level of Detail. District staff will use common sense and professional judgment to perform the appropriate level of detail of analyses to produce a quality project in a reasonable time and at a reasonable cost. Simplified evaluation procedures may be adopted for low risk/low cost projects and when the consequences of failure are minimal and do not pose a threat to human life or safety.

This reasonable policy should be applied more generally to Corps planning studies.

Reporting Results of Economic Analyses

Analytical results should be understandable to reviewers, to stakeholder groups, and to policy makers. Reports of analyses will be most understandable if they clearly state the assumptions used and models and data sources employed. At present, however, results from Corps planning studies tend to be presented in multiple, thick documents, and even the most diligent reader is challenged to identify the study's main assumptions and models employed, key environmental issues, primary costs and benefits, relevant risks and uncertainties, and so on. The evolution of environmental impact statements (EISs) done under the National Environmental Policy Act of 1969 offers useful lessons. After a chaotic period during the 1970s in which EISs became ever larger and more unreadable, reforms instituted by the Council on Environmental Quality in the early 1980s led to shorter, more consistent, and more usable documents, even for very large and complex projects. There is no reason the Corps cannot do the same. A short summary document should become a standard of Corps planning studies.

COMMENTARY

Economic concepts and analyses have a long history and tradition in the planning of Corps of Engineers projects, and economic concepts and considerations are at the foundation of the *Principles and Guidelines* and most of the Corps' investment and operational decisions. A review of the treatment of economics in Corps planning is thus in a sense a review of the agency's decision-making paradigms. Although economic concepts such as benefits and costs are clearly important in water resources management decisions, there is a danger in relying too heavily on monetized values and analytical methods to arrive at a final decision.

In Corps planning studies, the benefit-cost ratio, arrived at through analytical methods prescribed in the *Principles and Guidelines* and the *Planning Guidance Notebook*, is the ultimate criterion. This, however, is an improper use of benefit-cost analysis because it places an undue burden on analytical processes to determine the best water planning or management decision. Today, stakeholder groups call for a greater voice in water resources decision making, many important costs and benefits from Corps projects defy monetization and inclusion in the benefit-cost calculus, and political influences and preferences have always played important roles in water project construction and operations. Benefit-cost analysis clearly yields important information and should continue to be a standard in Corps planning studies. In fact, benefit-cost analyses in Corps planning studies should be strengthened by using accepted methods and external review procedures. However, a single measure, such as achieving the highest benefit-cost ratio, should not serve as the ultimate decision criterion. Implicit within the current planning process is an assumption that sound water management decisions are based primarily on monetized values derived by analytical methods, but in reality this is clearly not the case. Other important social, environmental, and even political considerations should be recognized explicitly in the decision-making process.

The *Principles and Guidelines* defines an analytical gauntlet that often requires years of experience to negotiate. Some degree of standardized procedures and guidelines is important to ensure consistency and efficiency, but the current *P&G* planning process may be inhibiting the exploration of more creative and comprehensive water management approaches. Moreover, planning guidance, models, and analytical methods developed since the 1983 approval of the *P&G* have not proven highly effective at helping resolve conflicts and differences between agencies, policies, or interest groups. The Corps of Engineers is not the only federal agency with problems related to overly detailed bureaucratic procedures, as pointed out in the 2003 Volcker Report: "Too often, as well, federal employees depart before their time in frustration over the strangling organizational and procedural complexity of contemporary government decision making. For too many, even their best efforts to be responsive and creative end up in organizational oblivion" (NCPS, 2003).

A review of economics methods in the Corps' primary planning guidance shows that these are not fully consistent with procedures used by other federal agencies and organizations. For example, the Corps is not allowed to count "damages avoided" as a benefit in relocating structures from floodplains, while in FEMA guidelines, damages avoided from relocations constitute a benefit. NOAA enlists "blue ribbon" expert panels and forges one set of valuation techniques and approaches, while the Corps adopts a different set. The USDA and the NSF have long-standing and vigorous external review programs, while the Corps has almost no tradition of routinely enlisting external experts in its planning studies. It appears that these agencies, despite many overlapping mission areas, nonetheless operate largely independently of one another. More vigorous efforts by the administration and the Congress to coordinate activities and responsibilities across federal water-related agencies would thus be useful. Such coordination would help the Corps and other agencies better understand each other's responsibilities, and could ensure that relevant, contemporary economics and other principles are being applied consistently across federal agencies.

Another point reinforcing the need for independent expert input relates to the fact that Corps staff positions are funded by individual studies or projects. This means that staff must charge their time to an active study or project, rather than being accounted for as general overhead. In addition, Corps district-level planning is conducted by staff members who live and work in the region under study and who may have personal relationships with the sponsor. Familiarity with the region and the sponsors may improve communication and a study team's sensitivity and appreciation for the priorities and values at hand, but it can also result in the Corps being placed in the position of acting as silent advocate rather than independent analyst. The motivation for the Corps to seek to please a local sponsor does not imply that this behavior is somehow improper, but rather speaks to the human nature of not wanting to disappoint colleagues and sponsors. Periodic, independent review can help ensure that these human tendencies are not affecting a study's engineering, economic, and scientific analyses.

In addition to weighing the input of independent experts, the Corps has for many years sought the advice of interested organizations and individuals in its planning studies and projects. Although the Corps has lengthy experience in working with "stakeholders" in its projects, finding an appropriate balance among diverse perspectives and values presents a substantial management challenge. Chapter 4 reviews the Corps' efforts to enhance stakeholder participation in its planning studies.

4

Stakeholder Participation

Stakeholder participation is an increasingly accepted component of natural resources and environmental planning processes in the United States and some parts of the world. In the U.S., stakeholder participation has been codified in environmental planning (e.g., the Administrative Procedure Act and the National Environmental Policy Act). Outside of the United States, international bodies such as the European Union, the World Bank, and World Commission on Dams have incorporated stakeholder participation into policy making and planning procedures. The importance of stakeholder participation in international water resources planning increased in the 1970s and continued through the 1990s and into 2000 and beyond (ESMPAP, 2003; IUCN, 2000; Mol, 2001; OECD, 1973; World Commission on Dams, 2000).

The Corps and other federal agencies have responded to changing forms of and demands for stakeholder participation in planning processes. Prior to World War II, federal agencies typically acted unilaterally, and public relations were oriented primarily toward gaining support for agency projects. After World War II, legal requirements of the 1946 Administrative Procedure Act stimulated more interest in public engagement, but this tended to be formal, highly structured, and oriented primarily to disseminate information. To facilitate opportunities for public comment, the 1966 Freedom of Information Act required full public disclosure of information. The National Environmental Policy Act of 1969 required preparation of environmental impact statements that are subject to public review and comment. During the 1970s the Corps increased its efforts to include stakeholder participation in planning. By 1978, the Corps had spent up to $80 million to facilitate public involvement, more than any other federal agency at the time (Rosenbaum, 1979). The Corps Institute for Water Resources was one of the first federal natural resource agencies to fund research and training in the field of public involvement (Creighton, 1983). In addition, the range of areas involving stakeholder participation within Corps studies was greater than in other federal agencies (Langton, 1993). In 1983, James Hanchey, then the Director of the Corps Institute for Water Resources stated, "For an engineering organization, public involvement has become crucial to our ability to provide engineering service to changing social values. Public involvement has helped define our role as engineers in the 1970s, and will continue to do so in the 1980s" (1988, p. 11). Box 4-1 discusses in further detail stakeholder participation in Corps planning studies.

ROLES IN DECISION MAKING

Although the notion of stakeholder participation continues to gain acceptance, there is no standard that constitutes "best practices" in this field. Because social, economic, and political conditions vary across settings, standardized, prescriptive stakeholder participation methods are likely to be inappropriate and ineffective. Stakeholder participation should be viewed more as a general principle applied in varied settings than a specific technique for wide application. One observer within the Corps explained this as follows:

> Public involvement is not a technique, but a strategy, approach or philosophy. There is no "one way" to do public involvement . . . . What works one place will not always work someplace else . . . . It is not the technique as much as the [attitude of the people] who employ the technique that is important (Delli Priscolli, 1993, p. 68).

Nonetheless, a variety of approaches and guidelines, whose effect will depend on the specific context and objectives, might be valuable.

Objectives of Stakeholder Participation

There are at least four objectives of stakeholder participation in planning:

1. Acceptance from the affected public. This objective is common to most planning efforts, but its meaning differs depending on how the planning process is implemented. In top-down planning processes, which are typically led by elected officials or planning "experts," the

BOX 4-1
Stakeholder Participation in Corps Project Planning

The Corps requires public involvement in its planning studies consistent with the Administrative Procedure Act of 1946 and the National Environmental Policy Act of 1969. The Corps has been a leader in promoting stakeholder participation, but practices at its district offices vary widely (for some good reasons), and the Corps has found it difficult to achieve consistent levels of stakeholder participation in project planning across districts. Also, expectations about stakeholder participation changed during the 1990s. Although it is now acknowledged that planning practices may vary according to context, contemporary expectations are that stakeholder participation should aspire to two-way communication, early and sustained public involvement, deliberations involving informal personal processes, and representation of all interests.

Four cases of contemporary Corps project planning were considered in this study in order to learn more about Corps planning procedures and public involvement in different districts (Chapter 1 of this report lists those case studies). All of these cases met Corps requirements for stakeholder participation, as defined in Corps planning guidance: (1) develop and implement an effective public involvement strategy as an integral part of the planning process for each study; (2) in cooperation with a non-federal sponsor, develop and implement a management structure to ensure effective collaboration in the feasibility study; (3) discuss how information gained from public sponsor involvement has been used in and influenced the planning process; (4) solicit comments on the draft report and environmental document. Beyond these requirements, the form, content, and extent of stakeholder participation are left to the planners' discretion.

Contemporary expectations for greater two-way interaction and direct public involvement in decision making are seldom met simply by fulfilling these requirements. For example, in the Houston-Galveston Navigation Study, concerns about channel deepening and widening were addressed openly and inclusively. Resource agencies and the public were involved through the evaluation period in multiple public scoping meetings and workshops. Coordination and cooperation among interested parties resulted in no major areas of controversy or unresolved environmental conflicts associated with the recommended plan. Committees of participating agencies and groups were assembled to identify concerns and to consider environmental impacts. Comprehensive studies were

Continues

Continued BOX 4-1

undertaken to assess potential impacts, alternative plans were developed, and recommendations were made and generally followed. This transparent process and inclusive stakeholder participation appeared to resolve some environment-related conflicts. Several factors may have contributed to this extensive stakeholder participation: Houston had a local sponsor committed to environmentally-responsible project implementation; natural resources agencies and user groups were more unified, vocal, and forceful in the Houston harbor expansion; and environmental concerns raised at each review level were acted on within the Houston-Galveston district and at the national level. Anecdotal evidence suggests, however, that participation does not always proceed in this fashion. The Corps would benefit from a review of stakeholder participation experiences and procedures across its districts. The Corps should also aim toward greater standardization in this realm, an effort that would be enhanced by the development of agency-wide manuals on stakeholder participation.

public is encouraged to support plans that have already been developed. Public officials and experts, however, today may be viewed with a greater degree of mistrust than in a previous era. In this context, efforts to get public acceptance might be seen as a cynical attempt by elites to co-opt the public (Mol, 2001). The public may be especially critical if it feels that it is being consulted only after fundamental decisions have been made. Frequent communication with the public in initial phases of planning is more likely to result in meaningful and satisfactory dialogue than if communications are enacted at later stages of the process.

2. *Managing risks and uncertainties*. All projects pose some degree of risks (e.g., risk of levee failure, risk that endangered species will not be protected, risk that forecast economic benefits will not materialize), and identifying and managing these risks is an important part of planning processes. Although experts play important roles in identifying hazards and levels of risks, they cannot determine levels of risk that are acceptable to the public and its various sub-groups. Ultimately, the decision regarding an acceptable level of risk is a public policy choice. The planning of complex projects also involves a degree of uncertainty about the range and magnitude of outcomes. This observation is especially true of large-scale projects involving significant environmental changes; such projects often offer large potential benefits, but the range and scale of

impacts are uncertain (Mol, 2001). The adaptive management approach is increasingly turned to in efforts to redress this dilemma in natural resources management, but even such projects typically cannot proceed without some willingness by the affected public to accept a degree of uncertainty (the 216 study panel on adaptive management also discusses stakeholder participation in Corps planning studies). Again, experts alone cannot and should not decide what uncertainties are acceptable or how to proceed in the face of uncertainty.

3. Education. The purpose of education is to create an informed public and thereby improve its abilities to participate in decision-making processes (NRC, 1999b; Popovic, 1993). Public education is part of a larger process that involves the development of public decision-making capacity with the intent of enhancing public involvement in decision making. Advocates of such education claim that it is needed because many contemporary planning problems entail a myriad of legal, environmental, and analytical complexities, whose details may be difficult for even the best-informed lay observer to comprehend. Efforts in public education are likely to be more effective if they are initiated in early stages of planning processes.

4. Building public consent consistent with democratic principles. This objective is based on the democratic norm that "citizens have rights to participate meaningfully in public decision making and to be informed about the bases from government decisions" (NRC, 1996). This objective perhaps represents the highest and most abstract standard for stakeholder participation. To be effective, it requires an educated public that is meaningfully engaged during the planning process. Stakeholder participation is meant to complement, not circumvent, political and decision-making processes. Stakeholder groups typically do not have the legal authority to set policy or resolve disputes. The inclusion of stakeholders is meant to help inform elected officials of the complex and dynamic interests of their constituencies.

Milestones in Public Participation Processes

Ideally, expert and public input would be integrated throughout the planning process, but one or the other may play a leading role at different times. Although their respective roles may vary through the planning process, useful stakeholder participation will engage decision makers and stakeholders at all planning stages. Key points in the planning process include the following:

- *Problem formulation.* At this stage, key questions are identified for further data gathering and analysis. To capture the complexity of a problem, scientific inputs as well as inputs from the affected parties are needed. Problem formulation should involve stakeholders, including the affected public, interest groups, agency officials, and scientific or technical experts.
- *Process design.* This involves establishing key parameters in the planning process, such as the appropriate scale that captures the full impact of a project and the full range of affected stakeholders (see the 216 study report on river basins and coastal systems for more detailed discussion of the issue of appropriate scales of planning). Experts can play important roles in identifying the appropriate scale of analysis by applying scientific theories to answer questions such as the following: What scale is best for addressing the problem or issue at hand? Can managers effectively influence critical areas, given the scale selected? Can the problem and the proposed solution be credibly evaluated? Is management at the selected scale feasible and economically affordable?
- *Selecting options and desired outcomes.* This is made possible by identifying trade-offs among alternative solutions and by identifying publicly held values that guide the selection of options. Scientific inputs provide a basis to identify trade-offs between alternatives. To understand the full range of trade-offs and alternatives, a variety of scientific inputs (ecological, physical, engineering, and behavioral) may be needed. Trade-offs may entail conflicts between fundamental and legitimate differences in values. It is especially important to recognize, respect, and deliberate on these points early in the study process.
- *Synthesis.* This attempts to reach consensus about a desired action. Recognizing that most decisions create winners and losers, it may be necessary to devise means to compensate those who suffer significant losses as a result of a particular course of action. At this stage, public input is central. Scientists play a role in identifying trade-offs, but the public should help articulate societal values during the decision-making process (NRC, 1999b). Stakeholder participation processes are imperfect and evolve in unpredictable ways, however, and they may not lead to the resolution of value-based differences and yield a single consensus. Such limitations should be recognized early in the process, and decision makers should be prepared to act in the face of differences and disputes. Despite its limitations, the inclusion of stakeholders in planning is a legitimate ideal in an open society. Given that stakeholder-based processes may not result in clear consensus, decision makers would benefit from a

synthesis report that explains the various points of view of stakeholder groups involved in a given plan. The Corps should include this type of synthesis in the summary document that this report recommends become a standard part of the agency's planning studies.

STAKEHOLDER PARTICIPATION APPROACHES

As mentioned, there is no standard for implementing stakeholder participation, but several approaches can be used singly or in combination to foster stakeholder participation. The execution of particular approaches—all of which have strengths and weaknesses (NRC, 1996, 1999b)—or the specific combination of them, depends on the issue or problem, the objective(s), and context in which they are employed. Examples of some common approaches include the following:

- *Public hearings*. This method may be the most common form of public engagement. It is called for in a variety of legislation, thereby meeting legal requirements for stakeholder participation. Public hearings allow for a variety of voices to be heard at one event; however, this form of engagement is often disappointing for several reasons. Most importantly, it is not clear it produces inputs that are actually incorporated into project planning. Public hearings are often held late in the planning process, and the forums tend to degenerate into posturing on the part of various special interest groups, resulting in little dialogue, mutual learning, or potential for consensus. Convening such forums early in the planning process may help demonstrate that stakeholder viewpoints are being used as part of the bases for fundamental planning decisions.
- *Citizen advisory committees and task forces*. Such bodies are typically appointed to address a specific issue for a limited term. These groups evaluate information over a period of time that allows for the development of a shared understanding of an issue or problem. The main limitation of such bodies is the question of whether they represent the broader public. For example, are all relevant interests represented, and are individual members effectively articulating the interests they represent?
- *Policy dialogues*. These encounters bring together stakeholders for the purpose of increasing understanding of a problem or issue. When not geared to the development of formal agreements, they help build common ground for future interactions. When geared to creating a formal agreement, these dialogues are referred to as "alternative dispute

resolution." These methods can be useful in situations involving sharp conflict and opposing opinions, beliefs, or values. They typically result in fewer protests and less litigation once a project is underway or completed. Like citizen advisory committees or task forces, the issue of representation is a concern (e.g., Who is involved and how were they selected?).

- *Surveys.* Surveys generate information about the knowledge, beliefs, values and opinions of a wide range of the public. If properly executed, they are effective at ascertaining the degree to which certain perspectives represent the broad views of the general population. Survey results, however, may be biased in terms of the content of the questions asked. If survey questions do not address relevant issues, the results may not be of interest or may represent a partial or biased view of the situation. Furthermore, surveys are simply a means of gathering information, not a means of promoting active deliberation about issues related to project planning.
- *Focus groups.* Focus groups are meetings of targeted subpopulations for concentrated discussion about a particular issue. These groups can help gather large amounts of information quickly with little expense. However, they share some of the shortcomings of the other methods. Like surveys, they are focused on gathering information and offer little opportunity for deliberation or group learning. Second, although focus group participants may be randomly selected, they are relatively small and the views presented in the discussions may not be representative of the general population or of targeted subpopulations.

Effective planning processes must integrate analysis based on scientific and technical inputs, with deliberation in a public forum that allows for meaningful participation of all affected stakeholders. Meaningful participation means that affected stakeholders deliberate about substantive issues related to any planning process, and that the results of this deliberation play a material role in project planning (NRC, 1996; 1999b). In one effort to codify general stakeholder collaboration guidelines, an NRC committee that reviewed Missouri River ecosystem science and the prospects for adaptive management offered a set of guiding principles (Box 4-2).

BOX 4-2
Stakeholder Involvement Principles
for Missouri River Dam and Reservoir System Management

- Participation by a broad spectrum of interest groups.
- Inclusion of tribal interests.
- Continuous two-way communication with the public.
- Visible participation by federal, state, and tribal governments and nongovernmental organizations.
- Support from an independent, interdisciplinary scientific panel.
- Provision by the federal government, with support from the states and tribes, of secure funding for stakeholder involvement efforts over the lifetime of the activity.
- Participation by representatives of Congress and of the state legislatures of the Missouri River basin states.
- Consensus decision making by the stakeholder group.
- Bounding the process with defined goals and with time lines for their achievement.
- Conduct of the activities of the governments in an open and transparent manner.
- Authentication of the stakeholder involvement process by governments in a formal document with all participating agencies as signatories.
- Provision of formal, independent facilitation for stakeholder group activities.

SOURCE: NRC (2002).

STATE OF CORPS PRACTICES

The overarching goal of public involvement for the Corps is to "open and maintain channels of communication with the public in order to give full consideration of public views and information in the planning process" (USACE, 2000; ER 105-2-100: 2-15). Although the Corps is careful not to relinquish its decision-making responsibility, the planning guidance lists four objectives of public involvement: (1) to provide information about proposed activities to the public; (2) to make the public's desires, needs, and concerns known to decision makers; (3) to provide for consultation with the public before decisions are reached; and (4) to consider the public's views in reaching decisions. The guidance recommends standard techniques for engaging the public, including information dissemination via the media, convening workshops and public meetings, and administering public opinion surveys. No specific criteria for the adequacy of public involvement are provided, but the expectation is that it be relevant and reflect the scope and complexity of the particular study. Corps guidance recommends that a public involvement strategy include several components that create opportunities for evaluation of stakeholder participation: a description of the preliminary consultation activities that led to development of the public involvement approach, including the agencies, groups, and individuals consulted; an identification of the public involvement expertise and effort that may be needed from various organizational units; and determination of the appropriate review points at which to evaluate the structure and function of the public involvement program.

These practices grew out of a legal and institutional foundation for stakeholder participation established in the Corps during the 1970s. However, as expectations for greater two-way interaction and direct public involvement in decision making grew, this framework has sometimes become an obstacle to deeper and more meaningful stakeholder participation. It has been noted, for example:

> NEPA's [the National Environmental Policy Act's] requirements . . . created a public participation model that results in dissatisfaction, frustration, and anger, because interest groups and individuals are viewed as data points; public involvement is treated as an analytical problem; and the decision is de-personalized. While studies evaluating the effectiveness of involvement techniques consistently show that participants prefer the informal, face-to-face techniques, NEPA and its implementing regulations mandate formalistic and impersonal approaches. Consequently, the available techniques tend to be one-way, while the most desirable techniques tend to be two-way (Cortner, 1993).

The promotion of stakeholder participation by federal agencies waned in the early 1980s. With a greater emphasis on local initiatives, federal agencies reduced the extent and intensity of efforts to promote public involvement. The centers for advancing public involvement shifted to the regional and local levels, where there continued to be demands for it. Appreciation of and demand for stakeholder participation grew within Corps district offices. A demand for public involvement was a clearly-expressed message in 16 regional listening sessions sponsored by Corps across the nation in 2000. Calls for stakeholder participation emerged in the discussion of a variety of topics including marine transportation systems, watershed management, project processes, and institutional changes. Several specific suggestions made at the listening sessions emphasized the demand for stakeholder or public involvement:

- Use problem-solving forums with all stakeholders to build consensus.
- Coordinate watershed planning involving all stakeholders and agencies (federal, state, and local).
- Emphasize full stakeholder involvement from a project's outset.
- Incorporate stakeholder inputs early in the process.
- Consider economic, social, and environmental benefits during project formulation.
- Increase interaction and communication with stakeholders.

Greater emphasis on locally initiated public involvement coincided with the Corps' requirement for greater cost-sharing in civil works projects pursuant to the 1986 Water Resources Development Act. Public involvement was required by this act, and public involvement activities could be counted as an in-kind, cost-shared contribution to projects by local sponsors (Langton, 1993).

As a result, the Corps decreased systematic consideration of stakeholder participation and it became more of a local or district matter. Today, local cosponsors often play a leading role in stakeholder participation. To the extent that the Corps takes less direct responsibility for stakeholder participation, its reviews are less structured, and expectations for Corps stakeholder participation activities have been reduced. As a result, there is no general sense of what the broader public's role is in the development of Corps projects, nor of how that public is represented. Instead, there is great variety across Corps districts and projects (Delli Priscolli, 2002, personal communication). Consistent with these observations in a previous NRC (1999a, p. 58) review of Corps planning procedures, it was concluded that:

> Current planning processes and funding arrangements have a tendency to force the Corps districts to view their constituencies narrowly, focusing upon the local sponsor. Efforts by local interests to include a broad range of participants in planning and to reach consensus

> on project plans require extra time, in some instances creating tensions between field level planners in the Corps and policy makers who are responding to the mandate to streamline the process. . . . The committee concluded that the local sponsor should be required to solicit the viewpoints of all interested stakeholders before asking the Corps to initiate a reconnaissance study.

The Corps has historically found it difficult to establish consistent levels of stakeholder participation in project planning across districts. Despite efforts in the 1970s to promote a broader range of and more open stakeholder participation practices at the district level, studies indicated that their adoption was limited. One study found that study directors were unlikely to use official public involvement guidance materials and only a handful of districts had established specific guidelines for their district (Langton, 1993). Another study reviewing the Corps' experience with stakeholder participation in the 1970s concluded, "Overall . . . the change in the decision process has been mixed at the field level, depending on both district initiatives and local demands" (Mazmanian and Nienaber, 1979). The authors went on to say, "The Corps is already doing better than most other federal agencies, even with its modest requirements for stakeholder participation in planning. We strongly suggest, however, that only outside pressure of the sort generated in the late 1960s and early 1970s will prompt the agency to institute an agency-wide open planning program . . . or seek dramatically different forms of public involvement (pp. 190-191)." These observations from the late 1970s might apply to the current state of affairs. The tendency to equate stakeholder participation with local project sponsorship potentially narrows stakeholder participation, but also introduces the possibility of considerable variation in the level and effectiveness of public involvement.

Systematic support and expectations for stakeholder participation (along with comprehensive, watershed scale planning) appears to be one outcome of local cost-sharing arrangements (NRC, 1999a). In instances in which watershed planning is conducted, however, stakeholder participation could be useful in cases where no single jurisdiction represents "the watershed." The lack of coincidence between many jurisdictional boundaries and watershed boundaries represents a long-standing challenge in water resources planning, and there have been a variety of government efforts to deal with it, including interstate compacts and commissions, the Tennessee Valley Authority, and various federal river basin commissions (Rieke and Kenney, 1997). In connection with this issue, Delli Priscoli (1993:64 1-64) observed:

> . . . public involvement is often viewed as a way of mobilizing a regionally affected constituency which cuts across state, local, and even international jurisdictional boundaries. By offering new opportunities for interested parties to interact, public involvement will encourage a broader spectrum of costs to be articulated, a more comprehensive trade-off analysis among alternatives, and increased regional plan acceptance by institutions and people within a region. Public involvement then becomes another strategy in the tradition of encouraging comprehensive and coordinated water resources planning.

COMMENTARY

Stakeholder participation represents a commitment to democratic principles in planning public works projects. A variety of contextually specific social and technical factors must be taken into account in engaging the public in project planning. Although planning practices will and should vary according to context, the stakeholder participation performance standards defined by the NRC Committee on Missouri River Ecosystem Science provide reasonable guidance toward establishing a set of general principles (see Box 4-2).

The Corps was an early leader in the development of stakeholder participation procedures and their implementation. Since the late 1980s, the Corps (like other federal agencies) has been less systematic in incorporating stakeholder participation into water project planning and management. Although it is required in all project planning, the design and conduct of stakeholder participation takes place at the district or local level with only very general standards grounded in legal requirements, such as those associated with the National Environmental Policy Act. Local cost-sharing and the potential to count public involvement as an in-kind cost-shared contribution create the potential for wide variation in the level and effectiveness of stakeholder participation. These activities may meet minimal requirements or may extend far beyond them, but the Corps does not have any means to assess the different efforts, or a system to identify and extend successful experiences to other projects. In the absence of a system of incentives or supports, practices may tend toward minimum standards that do not meet with contemporary standards for stakeholder participation.

Stakeholder participation processes defy standardization, evolve in hard-to-predict ways, and do not always yield consensus and clear direction. Especially in contentious situations, achieving consensus among different stakeholder groups may represent merely an ideal that is rarely achieved. Moreover, even if stakeholder groups were to achieve broad consensus, this outcome may be inconsistent with federal criteria for project approval. Yet as this chapter has explained, stakeholder involvement in project planning is essential and a potentially useful approach to implementing projects that reflect complex social and political differences and that can be adjusted to changing conditions following project implementation. Even agreement on small steps or objectives will provide some common ground for stakeholder collaboration, which can be useful in helping shape and refine objectives through the planning process. As stakeholder groups typically cannot fully resolve all differences, some synthesis of the different viewpoints would be useful to decision makers. These syntheses should be part of Corps planning studies. The Corps should also conduct a review of stakeholder participation procedures at its district-level offices to determine: (1) the balance between Corps and locally led stakeholder participation efforts, and (2) the level and effectiveness of stakeholder participation activities to create meaningful two-way communication between the public and Corps planners. The Corps should also prepare and publish training and reference materials on standards for stakeholder participation, which should provide useful, general guidance to planners throughout the agency.

5

Engineering

The Corps of Engineers played important roles throughout the twentieth century in developing and refining water resources engineering analytical approaches. There were major investments in research, both on a project level in Corps district offices and through programs of basic research at the Corps' five major research centers, which in 1999 were joined under a common administrative structure and are now collectively known as the *Engineer Research and Development Center* (ERDC; see Box 5-1).

The Corps is today known in the civil and environmental engineering profession for its development of a broad spectrum of analytical and modeling conventions. These contributions include the Corps' Hydrologic Engineering Center (HEC) family of hydraulic engineering computer codes, standards for in situ measurement of soil engineering properties, dredging and dredged material management guidance (including beneficial reuse options), structural models of massive concrete monolith construction, and artificial intelligence computer applications to support construction project managers. As a rule, practicing engineers look to the Corps for technological leadership. The Corps has responded by summarizing its design approaches in engineering manuals that have become standard texts that are used in universities and professional practice around the world, which are now available electronically through the Corps' Internet library.

The focus of this report is on Corps analytical methods and project evaluation approaches. In contemporary engineering practice, analytical methods and project evaluation are embodied in mathematical modeling. Although there are many dimensions to Corps engineering practices—from civil design, to hydraulics, to ecosystem intervention, to construction, and even to large-scale physical models of river reaches and coastal works—mathematical models represent the contemporary embodiment of analytical methods and, thus, the nexus of this chapter's discussions.

The Corps' traditional approach to engineering modeling—as, indeed, the practicing profession's approach more generally—has been to

BOX 5-1
Engineer Research and Development Center

The Corps operates five basic and applied research laboratories throughout the continental U.S. under the Engineer Research and Development Center. Each has a special mission, contributing to the breadth of science and technology needs demanded by water resources and military engineering projects. In order of size, these are the following:

1. Waterways Experiment Station, WES (Vicksburg, MS). The WES comprises several individual laboratories focusing on environmental, structural and geotechnical, coastal and hydraulic engineering, and information technology. The WES is the oldest and largest major Corps research facility, established in response to the Mississippi River flood of 1927. Started as a hydraulic modeling laboratory for river works studies, principally on the Mississippi and its tributaries, WES today undertakes broad research on both civil works and military engineering.

2. Construction Engineering Research Laboratory, CERL (Urbana, IL). Among the newer of the major Corps laboratories, CERL focuses on vertical construction applications, principally buildings and related facilities. The laboratory has been a leader in research on innovative materials for building construction, on construction technology and automation, and on advanced computer technology for construction and construction management.

3. Cold Regions Research and Engineering Laboratory, COREL (Hanover, NH). COREL has the specialized mission, reflected in its name, of developing and testing engineering technology for civil works and military engineering in cold climates, including polar regions. COREL conducts research on material and operations in winter battlefields; cold effects on construction; impacts of human activity on the environment of cold regions; and on the physics of snow, ice, and frozen ground.

isolate well-defined issues for analysis and to develop deterministic mathematical models or empirical design procedures by which to address them. Large factors of safety have been applied to calculated predictions of natural forces or to facility capacities in order to ensure exceptionally low likelihood that forces will exceed capacities and lead to failure. The result has been that hydraulic, structural, and similar facilities failures of Corps-designed works have indeed been rare. "Overdesign" of facilities, however, generally increases project costs.

4. Topographic Engineering Center, TEC (Fort Belvoir, VA). Unlike the other Corps laboratories, TEC has a predominantly military mission, developing and testing new technologies of mapping, spatial information sensing and processing, and battlefield information systems, and over the years has been closely related to the mission of the National Imaging and Mapping Agency (NIMA; formerly the Defense Mapping Agency). The TEC contributes to the Corps' civil works mission through its geographic information system technology and applications of remote sensing technology to terrain analysis.

5. Hydraulic Engineering Center, HEC (Davis, CA). HEC is a small research center dedicated to the development of hydraulic and hydrological engineering analysis methods and associated computer applications. Over the past four decades, HEC has developed computer applications for flood frequency forecasting, flood routing, sedimentation transport, flood hazard risk analysis, and other engineering problems, that have become industry standards in Corps projects and in the private sector.

The ERDC has been useful in many ways, offering specialized technical expertise not only to the Corps, but also to the academic and professional communities associated with various Corps activities. ERDC's engineering and design manuals have been internationally accepted and widely referenced for project guidance (e.g., ERDC's Shore Protection Manual, now being replaced by the Coastal Engineering Manual, is the internationally accepted design reference for coastal and navigation projects).

Historically, the Corps has had a close association with research institutions around the world and with the university research community, and has been known for its rational-analytical approach to engineering problems. Corps personnel have been active participants in professional societies such as the American Society of Civil Engineers, and its engineering methods have been widely aired at national conferences and in peer-reviewed archival journals. Historically, this has been in sharp contrast with the more empirical approach favored by much of the U.S. civil engineering profession (see Shallat, 1994), although the development of Corps technology, especially in earlier decades, has not been without its critics. Although traditional engineering models and approaches applied to planning generally produce reasonable results, efforts in improving algorithms, approaches, or enhanced applications should not be abandoned.

The primary role of engineering in decision making is in developing technical analyses and evaluations that are based largely on assumptions that aid in the eventual development of project alternatives. Engineering assumptions usually govern the project cost portion of benefit-cost analysis, which lends itself to close scrutiny, especially on controversial Corps projects. Although Corps design manuals have been an asset to the agency (and the engineering community in general), in some instances they can also act as limiting factors that discourage creative thinking by the Corps' project team (and project manager), thereby "boxing" them in to preset policy and procedural aspects. This has particularly been a problem in decision-making analysis, especially as technologies evolve and numerical techniques advance, with some Corps personnel still applying "old" techniques and technology. Although it can be argued that ERDC possesses state-of-the-art models, not all Corps districts have access to, or use those models. The Corps will have to make some adjustments in this realm if it is to possess high-quality technical expertise on engineering project analysis.

METHODS AND TECHNIQUES

Engineering models employed by the Corps on major projects can, as a first approximation, be categorized into four sets: (1) hydraulics and hydrology (H&H), (2) hydrodynamics and sediment transport, (3) geology and geotechnical, and (4) structural models. Other types of engineering models are used on civil works projects—for example, terrain models or environmental models of contaminant fate and transport—but these either are closely related to one or more of the four main categories or are of lesser importance in major project planning decisions.

Hydraulics and Hydrology

Hydraulic and hydrologic models treat the flow of water in natural and man-made channels or coastal regions, and the water cycle of rainfall, runoff, and groundwater. Hydraulic and hydrologic methods of analysis are used to make forecasts of flood frequencies (i.e., the probabilities that flood discharges of certain magnitude will be exceeded within given periods of time—for example, annually) or water heights at specific stream reaches with flood flows of known discharge values. The Corps has played an important historical role in developing these H&H methods. In order to support Corps planning in the future, it will be essential to develop models more fully, or perhaps develop model suites, linked to a common data base that more fully evaluate groundwater and surface water interactions, and that are capable of simulating a system for an extended period of time, rather than for a single rainfall event.

Hydrodynamics and Sediment Transport

Hydrodynamic models address the various intricacies of the coastal zone, including wave generation, development, propagation, and resulting processes. Sediment transport models are typically used hand-in-hand with hydrodynamic models to predict sedimentation patterns, littoral transport, beach stability, or near- and offshore sediment dynamics. The Corps has long been a leader in the development of scientific and engineering tools in this area, as well as in developing manuals. The recent publication in this series is the Coastal Engineering Manual (CEM), which will replace the Shore Protection Manual (SPM), also developed by the Corps, which has historically been used for the design and analysis of coastal and navigation engineering projects.

Geology and Geotechnical Engineering

These models treat the engineering properties and behavior of naturally occurring or treated geological materials—soil, rock masses, and groundwater—and the geological processes that affect those materials and structures constructed on or in them. Examples of such geological processes are earthquakes, regional subsidence, and swelling soils. Geotechnical models are used to develop forecasts of the strength of dam and levee embankments to the forces of water impounded behind them, the intensity of seismic ground shaking that the foundation of a structure can withstand without liquefying, settlement processes of nearshore and upland fills, or the spatial extent of contamination caused by groundwater seepage through a contaminated industrial site. The Geotechnical Engineering Laboratory at the Corps Waterway Experiment Station in Vicksburg, Mississippi, was an early and distinguished center for geotechnical research and has developed numerous design manuals and models for such applications.

Structural

Structural models treat the strength of structural systems, such as buildings, dams, breakwaters, bulkheads, jetties, and other constructed works, to withstand the loads—both natural and human-induced—to which they are subjected. Structural engineering is a broad and ancient practice, yet the Corps' unique purview of large water retaining and conducting facilities, as well as coastal protection structures, implies that the application areas it addresses are distinct from much of structural engineering practice associated with facilities such as buildings, roads, and bridges. Structural methods of analysis are used, for example, to evaluate the strength of large concrete dams, test the ability of a constructed breakwater to withstand extreme wave impacts, or test the dynamic responses of lock gates to varying loads of towboats. The Corps has developed several engineering manuals and design software of specific application for structural design, especially in the realm of coastal and navigational uses.

Systems Approaches and Perspectives

Water resources projects have become ever more complex and interconnected. For example, navigation, flood control, and ecosystem restoration projects on the Upper Mississippi River merge into one another geographically and create engineering, economic, and environmental impacts that cannot be separated from one another. Regional-scale projects, such as the Everglades Restoration, Coastal Louisiana, and CalFed, are becoming more common and involve complex interactions among impacts that have heretofore been primarily analyzed in isolation. These growing complexities necessitate a systems engineering approach to project planning.

A systems engineering approach to water resources planning involves a holistic view spatially, temporally, and across disciplines, which fundamentally changes the dynamics of planning. The systems approach has been driven in part by changes in technology, increasing pressure to lower costs and shorten project completion times, higher performance requirements, increasing complexity, and the increasing importance of information technology and real-time control in large civil works and environmental projects. This approach has also been driven by a growing awareness of long-range, distant-time, and interdisciplinary consequences of large projects that were not anticipated at the time of planning.

On the methods side, growing awareness of systems thinking in large water resource projects presents challenges to engineers and planners. These challenges include: (1) a growing awareness of a need to interrelate modular components of technology across disciplines in systems integrations; (2) a growing importance of teams of experts from different disciplines working together on complex projects, along with the associated problems of communication, interpretation, and documentation; (3) a centralization of information-dominated systems, which exploit commercial software and telecommunications technologies and derive information from a wide array of sources; and (4) a growth in volume of numerically intensive, multidimensional, heterogeneous, spatially distributed data that have to be accessed jointly by engineering, economic, and scientific planners.

These systems engineering issues are not unique to the Corps or even to water resources projects, and they arise in all aspects of large government and private sector projects. Given the Corps' traditional engineering and planning strengths and its national prominence, it is in position to aggressively pursue the development of systems engineering and economic planning methods that could benefit planning activities in natural resources management in the Corps and the federal government.

Risk and Uncertainty Analysis

The historical approach to coping with uncertainty in water resources engineering has been to design on the basis of a best-estimate of some extreme loading or stress (e.g., an extreme flood, extreme coastal storm, or extreme drought), and then augment that design by a fixed safety factor. The factor of safety is some multiple of the extreme event loading, for example, 50 or 100 percent more, or some other increment. This is an engineering tradition that is only now beginning to change in European and U.S. codes (Aashto, 1998; Cen, 1993). The Corps has been in the forefront of a transition from deterministic methods of analysis to risk-based methods that explicitly account for uncertainty (Box 5-2). This is an important transition because, as concluded by previous National Research Council committees (NRC, 1995, 1996, 2000) and detailed in Corps reports (USACE, 1992), risk and uncertainty-based methods lead to projects better tailored to local conditions and available data than those that are based on deterministic analyses. In particular, earlier deterministic approaches to project planning may not provide consistent levels of safety, consistent benefit-cost ratios, or consistent interpretations of environmental impacts across the nation. Such variations across projects and locales can, to a better degree, be captured by risk-based

BOX 5-2
Risk and Uncertainty in Flood Damage Reduction Studies

Many flood damage reduction projects involve the construction of levees. The historical approach to coping with hydraulic and hydrologic uncertainties of large floods was to base levee design on a best-estimate of the height required to retain a flood with an annual probability $p = 0.01$ of being exceeded (the so-called one hundred-year flood), and then augment that height by a standard levee "freeboard" of 3 feet. This became an engineering tradition across the nation.

Challenges to the concept of a standard increment of freeboard emerged in the early 1990s when it was noted that the standard 3 feet did not account for geographic or hydrologic differences at different locations, and thus afforded different levels of flood protection to different localities. This also drew into question the standard procedures for calculating the economic benefits conferred by levee freeboard.

The Corps' Hydraulic Engineering Center developed an innovative *risk analysis* approach to flood damage reduction analysis that holds great promise for rationalizing the way uncertainties are accounted for in project planning. This approach uses probabilistic methods, combined with statistical analysis of historical streamflows and stages and geographic information systems, to quantify the uncertainties associated with estimates of water heights and resulting property damage. The risk analysis approach has now become part of the National Flood Insurance Program levee certification procedure jointly conducted by the Corps and the Federal Emergency Management Agency.

SOURCE: NRC (2000).

methods. Even in common and traditional areas of project planning (such as construction cost estimation), the need for improved methods of capturing uncertainty is demonstrated by history (Shallat, 2002).

Although the Corps has been aggressive in developing risk-based methods of analysis for flood damage reduction studies and certain studies of navigation projects, methodological developments are needed (1) to strengthen the capability to employ risk-based methods on the broad range of engineering economic analyses used in project planning; (2) to extend risk-based methods to ecosystem restoration projects that involve large magnitudes of uncertainty; (3) to continue to expand the use of risk-based methods in traditional engineering disciplines such as structural, geotechnical, and coastal engineering; and (4) to enhance the development of uncertainty distributions about numerous other parameters.

Risk and uncertainty analysis can be a powerful tool if properly developed and constrained. The Corps should adopt a long-term focus toward enhancing and expanding the use of this analysis. The Corps, along with the U.S. Bureau of Reclamation, has taken a lead among engineering agencies in incorporating quantified expert judgment in risk-based methods of analysis (e.g., Klosterman and Sanders, 2000). This line of investigation should be continued, given the importance of engineering and scientific judgment to so many of the analyses used to evaluate water resource project plans.

Engineering Methods for Ecosystem Restoration

Hydraulics and hydrology and geotechnical methods of analysis are of central importance to the Corps' ecosystem restoration mission. Although ecosystems are biological, they depend ultimately on physical aspects of the natural environment, such as flow regimes, sedimentation patterns, and contaminant transport. Yet analytical methods that the Corps has traditionally used for hydraulic, hydrologic and geotechnical modeling were not developed with this use in mind and are in many ways inadequate to the needs of ecosystem restoration planning. For example, such methods are validated for making forecasts of relevance to a levee's required height to retain flood flows, but they are not (or at least not yet) validated for making forecasts related to, detention times in riparian wetlands. The latter may be critical to ecosystem restoration project planning.

Corps planners and engineers lack agency-sanctioned manuals that provide technical guidance and support for the design of engineered aspects of ecological restoration. Such manuals would be the equivalent of those now used for river hydraulic design, geotechnical site characterization, and structural analyses. They would refer to issues associated with the design, installation, and maintenance of systems that make use of natural materials; to the size and strength of a structure used within the restored area; to engineering adaptations affecting fluvial geomorphology; and to other engineering methods that lend themselves to treatment in a technical manual. Corps designs are generally influenced by a tradition of protecting people and property, where the consequences of failure are costly to human health and safety; in contrast, in restored areas the consequences of failure may be less dramatic. In part, the lack of manuals may result in a lack of willingness by more cautious designers to apply engineering techniques of ecosystem restoration or may lead to "overdesign" and increased costs.

COMMENTARY

Three aspects of the Corps' engineering analysis and methods bear close attention in the years ahead. The importance of these relates to the changing paradigms of U.S. water resources management and to the changing needs of Corps projects and activities. These are (1) systems engineering aspects of water resource planning, (2) impacts of risk and uncertainty on planning, and (3) integrating engineering methods of analysis with ecosystem restoration planning. Another important consideration in Corps planning studies, and of the engineering methods therein, is the value of occasional independent review. Corps planning studies would also benefit by including a summary of the key assumptions used in engineering design, models, and methods of analysis. These dimensions of a planning study are often extremely complicated and technical and are difficult for the layperson to understand. Furthermore, they often tend to be located at various places in a planning study, making it difficult to grasp the key issues and approaches quickly.

Equally important to ensuring the use of credible methods and techniques is having the resources to apply and implement them. Although the Corps may have some ability to stay abreast of engineering and technical advances, it suffers from a limited ability to recruit and retain talented personnel. The Volcker Report looked carefully at the issue of personnel within the federal government. Its observations parallel some of this report's observations regarding Corps personnel issues (NCPS, 2003):

> Far too many talented public servants are abandoning the middle levels of government, and too many of the best recruits are rethinking their commitment, either because they are fed up with the constraints of outmoded personnel systems and unmet expectations for advancement or simply lured away by the substantial difference between public and private sector salaries in many areas.

Support from the administration and Congress for the Corps to recruit and retain well-qualified staff, to be able to hire staff from outside the agency, and to create realistic and rewarding career advancement paths, will all be important to the Corps as it addresses twenty-first century engineering and planning challenges.

6

A New National Water Management Framework

For much of U.S. history, water resources were viewed as a means for promoting economic expansion. Until the middle of the twentieth century, Corps of Engineers projects were generally single-purpose, often built primarily to manage flood risks or to enhance navigation. The Corps executed its work program with clear lines of authority from the U.S. Congress and with widespread citizen support. During the latter half of the twentieth century, the Corps began to implement multiple-purpose projects that aimed to provide a broader suite of benefits. This was a period during which preferences for the benefits sought from river, wetland, and coastal systems were broadening. Increasing interest in conservation, greater demand for water-based recreation, and the loss of some pristine riverine and coastal areas conspired to change the services sought from Corps projects. Water projects constructed in the middle of the century for hydroelectric power generation and flood control, for example, today may also be used to support flat-water recreation or to provide instream flows. Furthermore, restoring varying degrees of flow regimes and enhancing ecological benefits are likely to be prominent components of the Corps' future work program.

These shifts have had profound consequences for Corps project planning and management activities. For example, shifting environmental perceptions and preferences in the United States starting in the 1960s gave rise to federal legislation aimed at environmental protection. The National Environmental Policy Act (1969), Clean Water Act (1972), and Endangered Species Act (1973) are examples of federal environmental legislation with which the Corps must comply. In the ensuing decades, additional legislation, mandates, and tasks cited in congressional committee language have piled up. As a result, the Corps of Engineers today must comply with more than 200 pieces of federal legislation, some dating back to the early nineteenth century. This multilayered body of legislation was not created in accord with a master plan or strategy, and it has accreted on top of existing legislation that generally has not been rescinded. It thus contains inconsistencies and does not provide guidance on resolving conflicts between mandates. Moreover, mandates to the Corps are often cast in broad, qualitative terms that do not offer specific guidance on the means of attaining desired ends or even describe when an objective has been met. At one time, this lack of specificity resulted in few problems because the Corps' expertise was taken for granted, decisions were not widely scrutinized, Corps projects did not entail controversial trade-offs, and citizens generally agreed on national water management objectives. The contemporary setting, however, is vastly different, and involves scrutiny and criticism of national water management plans, a broad array of values (some of them mutually exclusive), increased willingness to litigate, and increasingly complex problems. A lack of clarity regarding the relative importance of legislation and guidance and a lack of specificity about desired objectives inhibit the agency's ability to reach clear feasibility study recommendations.

These shifts have had important consequences for Corps planning guidance. The federal *Principles and Guidelines (P&G)* document is geared to identify a project planning alternative that maximizes net economic benefits (the national economic development, or "NED," alternative). In a contemporary political setting of broadened social values and lack of broad consensus about whether water resources should be used primarily to promote economic development or also to promote environmental or social benefits, a focus on traditional NED planning alternative may no longer be appropriate. Many U.S. citizens today perceive water resources as having value beyond an ability to provide for economic growth. Yet values associated with environmental preservation and aesthetics are not as easily monetized as traditional economic values associated with navigation and flood control and, thus, are not easily captured in the *P&G* analytical framework. Moreover, the *Principles and Guidelines* document was framed on the assumption that water resources planning decisions are primarily analytical in nature. In a "progressive era" setting (see Hays, 1959, for a discussion of late nineteenth-early twentieth century rational planning approaches) in which experts were entrusted with planning decisions and decisions were based on engineering principles, the *P&G* framework might have been appropriate. However, in a setting of conflicting values in which the need for stakeholder participation is broadly accepted, and where problems are complex and fraught with uncertainty, the *P&G* analytical framework is less able to effectively guide water resources management decisions.

CHANGES IN THE FEDERAL WATER POLICY FRAMEWORK

Resolving Discrepancies in National Water Policy

The Corps is hindered in its ability to reach clear planning and management decisions because of inconsistencies and contradictions that exist in the body of legislation, congressional committee language, and administration guidance (budget instructions, executive orders, and Office of Management and Budget guidelines) that constitute *de facto* national water policy. For example, the Corps of Engineers is obliged to follow the ESA and the CWA, but it is also authorized to maintain 9-foot navigation channels on river systems such as the Missouri and Mississippi. These mandates, however, are not always fully compatible. For example, a federally endangered species of fish may depend on some degree of streamflow variability at critical stages of its life cycle. The goal of protecting this species would thus, at times, be at odds with a goal of providing a reliable navigation channel. Such issues are further complicated by significant scientific uncertainties regarding the dynamics of large river-floodplain ecosystems or in determining the needs of a particular species. Without clear guidance on how to resolve such policy conflicts, the Corps often muddles through with a status quo plan and is then criticized for its inability to identify and implement new management decisions. A National Research Council committee that reviewed science and decision making on the Missouri River made a similar observation and recommended that "support of the U.S. Congress is ultimately needed to help establish acceptable goals for the use and management of the Missouri River system. Congress must also help identify the necessary authorities to do so" (NRC, 2002a). Similar examples can be found in conflicts among legislative instructions, agency regulations, and administration guidance concerning floodplain development, environmental enhancement programs, economic analyses, and other water issues.

The lack of clarity and consistency within the body of federal water policy hinders the Corps' ability to develop contemporary methods and models for use in its planning studies. Corps planners in the agency's district level offices do not direct their efforts toward meeting a set of clear, internally-consistent national water policy goals, but rather focus on 1) adhering to steps prescribed in the *Principles and Guidelines* and related planning guidance, and 2) identifying a NED alternative that has a benefit-cost ratio greater than 1. Thus, rather than being provided with clear ends and with a charge to develop means for achieving them, Corps planners are typically provided ambiguous ends, but with a set of relatively rigid steps to be executed. These steps—embodied in the *P&G* and in Corps planning guidance—stymie approaches or models that stray too far from the guidance and minimize the prospects for using the local knowledge and creativity of Corps planners to address water-related problems. Further complicating this setting is that the Corps often works in partnerships with local cosponsors who cover some portion of a planning study's costs. The local cosponsor often has a clear idea of its desired planning alternative and has little interest in exploring alternatives or in supporting studies or projects that promote the national—as opposed to local—interest. With a lack of clear federal-level guidance or objectives, and a local sponsor that knows what it wants, that sponsor's opinion may carry the day. Local stakeholder input should be included in management decisions, especially if local cosponsors are responsible for some portion of the study costs. However, the Corps of Engineers is a federal agency whose primary responsibility should be to uphold the public interest.

Another example of a lack of clarity and consistency in legislation that guides the Corps is found in the 1983 *P&G* and in subsequent Water Resources Development Acts. The federal objective as defined by the 1983 *Principles and Guidelines* is "to contribute to national economic development, consistent with protecting the nation's environment, pursuant to national environmental statutes, applicable executive orders, and other Federal planning requirements" (WRC, 1983). This objective, however, fails to require equivalent consideration of other factors as mandated by the Water Resources and Development Acts of 1986, 1990, and 1996 (codified in 33 U.S.C. section 2281). Section 2281 states in its entirety:

> Section 2281. Matters to be addressed in planning. Enhancing national economic development (including benefits to particular regions of the Nation not involving the transfer of economic activity to such regions from other regions), the quality of the total environment (including preservation and enhancement of the environment), the well-being of the people of the United States, the prevention of loss of life, and the preservation of cultural and historical values shall be addressed in the formulation and evaluation of water resources projects to be carried out by the Secretary, and the associated benefits and costs, both quantifiable and unquantifiable,

> and information regarding potential loss of human life that may be associated with flooding and coastal storm events, shall be displayed in the benefits and costs of such projects.

This directive mandates that all these factors be addressed along with national economic development in the Secretary's evaluation of water resources projects, and that any cost-benefit analysis reflect consideration of these factors, whether quantifiable or unquantifiable.

As noted, national water policy consists of a large number of legislative instructions and congressional and administration guidance. This amalgamation of policies and direction, however, does not always provide clear guidance or management objectives and often exhibits internal inconsistencies. **To provide clearer direction to the Corps, the administration, and the Congress, in cooperation with the states, should reconcile inconsistencies within the existing, de facto, body of national water policy (Recommendation 1).**

Coordinating National Water Policy

More than a dozen federal departments and independent agencies are responsible for some dimension of water resources management, and for the most part, they operate as independent activities without administration coordination or water policy oversight. Their charters come from the directives of the OMB, from legislative guidance, or from instructions of Congress. The principal water resources management agencies (the Corps, the Bureau of Reclamation, the Environmental Protection Agency, the Natural Resources Conservation Service, and the Tennessee Valley Authority) must coordinate and interact with resource and science-based agencies (the Fish and Wildlife Service, the Forest Service, the National Park Service, and the National Oceanic and Atmospheric Administration Fisheries), as well as specialized agencies with other water-related responsibilities (e.g., the Department of Housing and Urban Development, Department of Homeland Security [Federal Emergency Management Agency, Coast Guard], and the Small Business Administration). They must also work with the Executive Office of the President (the Council of Economic Advisers, the Council of Environmental Quality, and the OMB). Many issues are identified and resolved among or between agencies by senior career officials through memoranda of agreement and similar instruments; however, more complicated issues—those that would require participation of political appointees at the assistant secretary level or higher—are frequently shunted aside because of the difficulty of bringing these individuals together. Even when meetings within multiple agencies do take place, the results are frequently bounded by the respective agencies' legislative authorities, may not jibe with fiscal realities or administration policies, or may not be made known to other water-related agencies. For example, standards for levee design, cost-benefit analysis, floodplain occupancy, and postdisaster recovery support vary among agencies. Differences in methods and standards between agencies with overlapping or related responsibilities can impede effective program and policy execution (see NRC, 2000).

The federal Water Resources Council (WRC) formerly provided a formal mechanism for promoting inter-agency coordination. Created as part of the 1965 federal Water Resources Planning Act, the Water Resources Council consisted of the Secretary of Agriculture, the Secretary of the Army, the Secretary of Commerce, the Secretary of Energy, the Secretary of Housing and Urban Development, the Secretary of the Interior, the Secretary of Transportation, and the Administrator of the Environmental Protection Agency (the latter was added after the EPA was established in 1970). Representing a culmination of many decades of interest in comprehensive river basin planning, the Water Resources Council was formed to report to the president and was charged with establishing principles, standards, and procedures for water resources and related land use planning (Kneese, 1996). The WRC established four different objectives for planning: (1) Regional Economic Development (RED), (2) Environmental Quality (EQ), (3) Other Social Effects (OSE), and (4) National Economic Development (NED). When the Water Resources Council issued the 1973 *Principles and Standards*, however, actual planning procedures were listed for only EQ and NED. In addition to the *P&S*, the Water Resources Council also issued assessments of the nation's water resources, as well as a design for a state-level planning program. The council drew a great deal of criticism, however, and it was zero-funded in 1981. The Water Resources Council may technically continue to exist, but only on paper, as it has no staff and convenes no meetings.

Whatever limitations existed within the Water Resources Council, it was the only venue for formal discussing and coordinating issues of import to the federal water resources management agencies. The functions lost with the demise of the Water Resources Council, such as its promotion of formal inter-agency coordination, have not been restored. As a result, administration-level coordination meetings over the past two decades have been infrequent, and loose ends and conflicts abound. Plans to address water quality issues may not be closely aligned with water quantity-related issues or environmental activities. Interagency discussions are *ad hoc*. The development and application of new and innovative planning methods—even those that may have important inter-agency implications—are not formally coordinated between agencies (see, for example, NRC, 2000). This lack of formal, routine, and high-level interagency cooperation acts as a barrier to consistent policy applications, proactive and innovative planning approaches, and to the development and standardization of planning methods and techniques. Although the Water Resources Council was not without its limitations, the loss of the council has contributed to policy fragmentation and to inefficient and inconsistent applications of planning approaches and methods. **A body should be specifically charged to coordinate water resources policies and activities among the administration, the Congress, the states, and federal water resources management agencies with water resources management responsibilities (Recommendation 2).**

Revising the *Principles and Guidelines*

Policy Dimensions—Principles

The Corps' planning procedures are governed by the federal *Economic and Environmental Principles and Guidelines for Water and Related Land Resources Implementation Studies* (the *P&G*). Written in 1983 by the federal Water Resources Council (WRC), the *P&G* also guides water resources project planning of three other federal agencies: the Bureau of Reclamation, the Natural Resources Conservation Service, and the Tennessee Valley Authority. The "principles" within the P&G identify the objective of federal water resources activities as: "To contribute to national economic development consistent with protecting the Nation's environment, pursuant to national environmental statutes, applicable executive orders, and other Federal planning requirements" (WRC, 1983).

This national economic development objective represents the views of the current and previous administrations, and not necessarily those of the Congress. The emphasis on economic development discounts the value of environmental and social costs and benefits. Moreover, the

P&G outlines a process that causes costs and benefits to be represented only by what can be monetized, which does not allow for full consideration of a project's nonmonetized aspects. The *P&G* therefore directs the Corps (and other agencies) to develop projects that do not reflect full pricing of the costs and benefits of federal activities. This is especially problematic in that a) nearly every Corps of Engineers prospective or existing water project has important (and difficult-to-quantify) environmental dimensions, and b) that an increasing portion of Corps projects are designed to enhance environmental benefits.

This federal objective and the procedures outlined in *Principles and Guidelines* for the execution of project planning do not incorporate either policies and statutory objectives that recognize social needs and preferences or planning approaches and methods that have evolved since 1983. This results in conflicts among guidance found in the *P&G*, statutes enacted since 1983, and contemporary approaches to project evaluation. The Corps is thus torn between administration direction to conform with the *P&G,* and administration or congressional guidance to support other federal objectives and the polices and procedures found in various legislation and in other administration and congressional documents. Moreover, other federal agencies conducting important water planning and management activities (e.g., the Environmental Protection Agency and the Federal Emergency Management Agency) are not required to follow the same project justification procedures, despite the fact that their impacts on water resources may in some instances exceed those of the Corps.

Analytical Dimensions—Guidelines

The guidelines in the *P&G* are used to identify the national economic development alternative. Alternatives to the NED may be formulated, but they are not required. In addition, all alternatives are subjected to a benefit-cost test. The guidelines prescribe steps to be followed in the planning of a variety of water-related projects, including municipal and water supply, agriculture, urban flood damage, hydropower, inland navigation, transportation (deep-draft navigation), recreation, and commercial fishing.

There is a strong reliance within the *P&G* planning guidelines on forecasts and future estimates. Figure 6-1, for example, shows the steps prescribed in the *P&G* for planning an urban flood damage reduction project. This figure illustrates that the planning process and final deci-

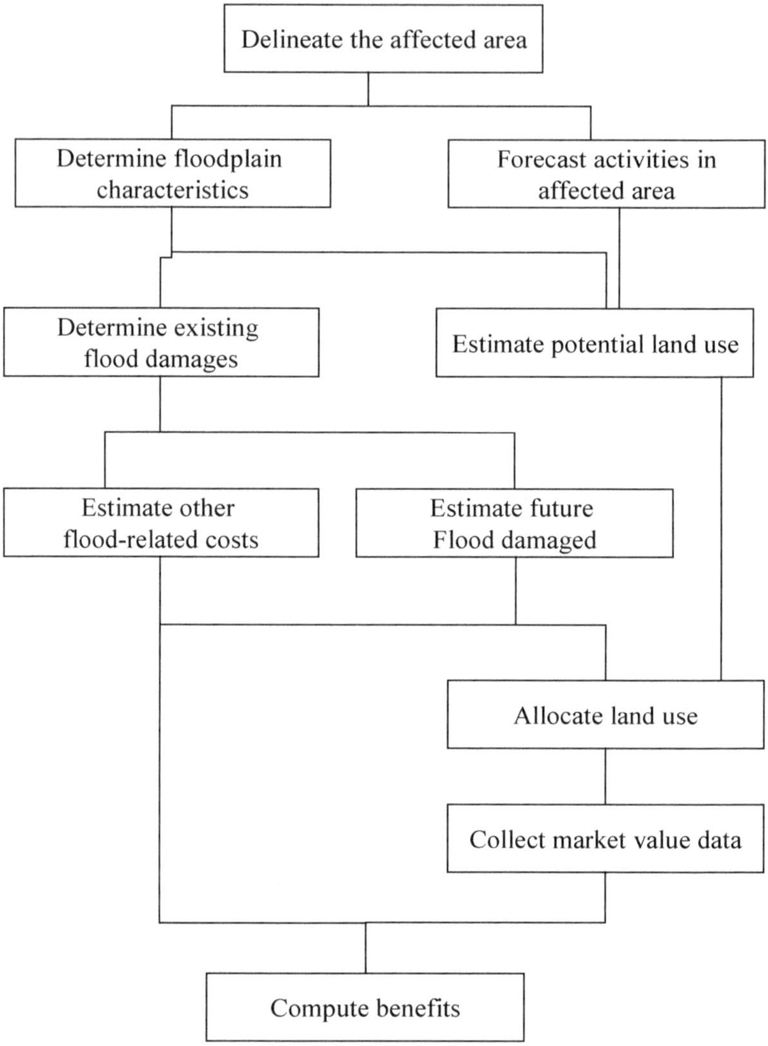

FIGURE 6-1 Steps in an urban flood damage reduction study, according to the *P&G*. SOURCE: WRC (1983).

sion depend on forecasts and future estimates of activities in the affected area, land use, flood-related costs, and future flood damages. Within the guidelines, 10 steps are prescribed for evaluating urban flood damage projects, many of which call for the calculation of values for multiple variables. Step 7 (WRC, 1983, p. 2.4.11), for example, calls for the computation of hydrologic changes, economic changes, projection of physical damages, physical units, value per physical unit (existing, future, translation to flood damages, limit), and damage susceptibility (projection of income losses and projection of emergency costs). Not only does the calculation of some of these variables—including many future and thereby unknown variables—require a degree of analytical sophistication, but their inclusion and quantification as integral planning steps are based on the premise that they can be predicted with a reasonable degree of precision. This example not only illustrates the degree to which the *P&G* relies upon precise and accurate forecasts of future variables, but also the great level of detail contained in the guidelines, which amount to more than 100 pages of instruction.

In addition, many concepts and approaches in the *P&G* were derived from its predecessor *Principles and Standards* document (WRC, 1972), which was based on paradigms that dated back to the 1950s and 1960s. Although some of the conceptual foundations within the *P&G* may yet have relevance, the document does not reflect analytical advances (e.g., monetization techniques, risk and uncertainty analysis, changing views of appropriate benefit-cost applications) and shifts in planning paradigms (e.g., adaptive management, stakeholder collaboration, independent review) that have occurred since 1983.

Another key consideration in contemporary water resources planning not adequately reflected in the *P&G* is stakeholder participation. In today's planning environment, interest groups from all corners carefully scrutinize all aspects of Corps planning studies. To its credit, the Corps generally promotes strong public involvement in its studies, and the Corps routinely holds public meetings at various stages of its planning studies, especially for the agency's more visible and controversial studies. Informal discussions with Corps planners and analysts suggest that staff members appreciate the value of early public participation and encourage it. The *P&G* analytical framework, however, provides no avenue for the input or participation of stakeholder groups. The *P&G* contains a short paragraph on "General Public Participation" and appropriately recommends participation "throughout the planning process" of national, regional, local, and tribal groups. The analytical procedures prescribed in the *P&G*, however, do not allow for qualitative considerations to enter into the planning calculus. The *P&G* thus puts the Corps in a position of working in a setting in which stakeholder groups request input into planning decisions (which the Corps appreciates and responds to), but of having to follow a step-by-step framework that does not allow for the consideration of nonquantified variables.

The *P&G* represented state-of-the-art thinking in economics and planning when it was written and some of the concepts and paradigms that underpin the *P&G* are relevant today. However, it has been over 20 years since the *P&G* was updated and revised. The *P&G* planning process is part of a federal water resources decision making environment in which the product has become adherence to process, rather than sound projects and operations plans that serve contemporary needs.

The Principles and Guidelines document should be revised to better reflect contemporary management paradigms; analytical methods; legislative directives; and social, economic, and political realities. The new planning guidance should apply to water resources implementation studies and similar evaluations carried out by all federal agencies. A revised version of the P&G document should be periodically and formally reviewed and updated (Recommendation 3).

No significant action has yet taken place within the Administration in response to this recommendation that has been voiced by many groups. **Therefore, even if the Administration should choose to not revise the P&G, the Corps should draft a revision to its Planning Guidance Notebook that is consistent with this report's recommendations, and propose this revision to the Administration (Recommendation 4).**

7

Revising Corps of Engineers Planning Studies

BENEFIT-COST ANALYSIS AND FEASIBILITY STUDIES

One limitation of the federal *Principles and Guidelines* is its reliance on the benefit-cost test as the key criterion for most water resources planning decisions (ecosystem restoration projects, for example, are not subjected to this criterion). The *P&G* is based on the presumption that all important costs and benefits related to a planning decision can be quantified and monetized, and that the computation of such should be (and will be used) to determine the fitness of a planning alternative. The *P&G* thus presumes that water resources planning decisions are largely analytical in nature, as social and cultural issues, for example, are not included as components in the *P&G* evaluation process.

This approach does not adequately reflect the reality of current analytical paradigms or of today's planning and political realities. Economists have achieved significant advances in monetization techniques since 1983, through the use of travel cost, contingent valuation, and other methods. But economists also acknowledge the limits of such approaches and recognize that public policy decisions should be based on considerations and analyses beyond merely comparing total benefits and costs (see Arrow et al., 1996, as discussed in Chapter 3). Examples of these factors include non-monetized values, uncertainty in forecasts, and equity concerns.

Benefit-cost analysis should not be used as the lone criterion in deciding whether a proposed planning or management alternative in a Corps planning study should be approved (Recommendation 5). A more appropriate role for benefit-cost analysis is to serve as a primary source of information concerning the benefits and costs of project alternatives, and the groups who gain most from a project. This separation of the role of benefit-cost analysis from its use as a mechanistic decision criterion would reduce the pressure on Corps analysts to seek a high degree of precision, which does not always reflect a similar degree of accuracy. It would also relieve the pressure placed upon the *P&G* document to produce a single and credible numeric figure that no planning framework is capable of reliably producing. Benefit-cost analysis is a useful method for promoting sound public policy and it should continue to be a part of Corps planning studies. However, these benefit-cost analyses in future Corps planning studies would ideally contain clear and concise summary statements, explicitly and succinctly explain key assumptions and models, and be subjected to reviews of independent experts.

REVIEW OF PROJECTS AND PLANNING STUDIES

Ex Post Studies

For much of its history, the Corps used primarily engineering techniques to construct civil works projects. As mentioned, desired project ends were clear and were often single-purpose. Expected project benefits, such as flood control or navigation enhancement, were usually immediate and visible and there were thus few questions about a project's effectiveness. In this setting, the Corps planned and constructed projects and moved on to the next project (except when maintaining a presence at some projects for operations and maintenance purposes). The idea of revisiting Corps of Engineers projects to identify how well they performed, and using these lessons to improve future planning and management, was not widely considered. Moreover, water resources and other agencies tend to resist reviews of past projects for several reasons, including a preference to allocate resources for actual construction rather than investigations, and a reluctance to have past mistakes identified. For nearly fifty years, one of the nation's eminent water resources experts, Gilbert White, has pointed to limited evaluations of project results as a key water management shortcoming:

> We could fill a large room with documents drawing up what are considered the best plans for an analysis of problems in river basins around the world . . . On the other hand, the literature about what has happened after any of the projects have been carried out can be assembled on one end of a small table. There is no tradition of making retrospective or evaluative studies of the consequences. For example, no evaluation of the Tennessee Valley Authority has been undertaken . . . we have no satisfactory explanation of why, thirty years after the

> TVA was started in order to develop the economy of the region, the major part of the Tennessee Valley is still considered an underprivileged area of Appalachia, deserving special subsidy contributions from the federal government for further improvement (White, 1971).

Corps projects today are often operated in very complex legal and social settings, projects are expected to meet the needs of multiple users with shifting preferences, and there is wider recognition of economic and environmental uncertainties associated with water projects. That water resources projects often have a range of unintended consequences, and that operational schemes are likely to require re-adjustments, is also better appreciated. These issues have given rise to planning concepts such as "adaptive management," which recognizes uncertainties and emphasizes careful monitoring and evaluation of environmental and related outcomes to promote flexible resource management policies that can be adjusted in a changing and unknown future world. The Corps has planning authorities that allow for project operations to be reviewed and adjusted. The two authorities that the Corps uses most frequently for these purposes are a "Section 216" authority from the 1970 Flood Control Act and a "Section 1135" authority from the 1986 Water Resources Development Act. For example, the Corps is conducting its current feasibility study of the Upper Mississippi River-Illinois Waterway under its 1970 Section 216 authority (USACE, 2002b), while the Section 1135 authority is directed more specifically toward ecological restoration projects. Although these authorities allow the Corps to conduct ex post studies, these authorities do not direct the Corps to do so, nor do they appropriate resources for their conduct (see the 216 study report from the coordinating committee for further discussion on a new study authority for the Corps). Better management of existing infrastructure will require more frequent and more extensive reviews of the ecological and environmental outcomes of existing Corps projects (see the 216 study report from the panel on adaptive management for more discussion on post-construction monitoring). A better understanding of the environmental, economic, and social outcomes of Corps projects is essential in helping the agency learn from past successes and failures and to provide information for helping formulate new operations plans. **Periodic reviews of completed projects should be a routine part of Corps project planning and management. Congress should provide resources to conduct these ex post evaluations (Recommendation 6).**

A Strengthened Reconnaissance Study

Corps of Engineers water resources project planning studies are conducted in two stages—a reconnaissance stage and a feasibility stage. The reconnaissance study is conducted to determine if there is a federal interest in a given water resources opportunity or problem. The Corps' internal guidelines limit reconnaissance studies to one year and limit those studies, which are fully federally-funded, to a cost of no more than $100,000. The large majority—roughly 6 of 7—of Corps planning studies do not proceed beyond a reconnaissance study.

The cost and time constraints placed upon reconnaissance studies have several implications. A possible strength of limited reconnaissance studies is that they may be useful in identifying and dismissing non-viable proposed projects for a modest amount of resources. That is, if reconnaissance studies are used primarily to "weed out" sub-par proposals, these limits may be reasonable. On the other hand, these limits may not allow for careful consideration of a broad range of planning or management alternatives, especially with the Corps' large projects, more controversial projects, or with projects operated in a systems framework (e.g., the Missouri River dam and reservoir system). These limits place pressure on Corps district-level planners and analysts to make a relatively quick decision regarding a given water resources issue. But hastily-conducted reconnaissance studies may not allow the Corps to adequately consider options for addressing a given problem, a problem that may be exacerbated by a local co-sponsor who wants to move quickly through study reconnaissance. The Corps' role in reconnaissance studies should be to objectively consider a water resources problem in light of the possible alternatives, something that may not always be possible within the one year/$100,000 limits. Moreover, although one year and $100,000 may be adequate for a reconnaissance study of a local project, these limits are inadequate for studies such as the Upper Mississippi River—Illinois Waterway system. The reconnaissance phase of Corps of Engineers reconnaissance studies should be conducted with an expectation that the Corps will identify and examine a broad range of alternatives in order to provide conclusive evidence about the federal interest in a water resources issue. **Resources and time allocated for Corps reconnaissance studies should be commensurate with the scale and complexity of the water resources issue at hand (Recommendation 7).**

STAKEHOLDER PARTICIPATION

The Corps played a leading role in the development of stakeholder involvement/public participation practices. At the same time, it has long been observed that the quality of such practices is inconsistent across Corps projects. Concerns about the unevenness of practices have re-emerged with the advent of local cost sharing and a high degree of stakeholder participation. Furthermore, stakeholder involvement has evolved significantly over the past thirty years and the current state-of-the-art emphasizes meaningful two-way communication between stakeholders and project planners.

A comprehensive review of stakeholder involvement/public participation procedures at the district level would allow the Corps to determine: a) the balance between Corps and locally-led public participation efforts; b) the level and effectiveness of public participation activities to create meaningful two-way communication between the public and Corps planners. This will allow the Corps to identify opportunities for more effective stakeholder involvement and public participation and to re-assert its leadership in promoting state-of-the-art practices among natural resources and environmental management agencies.

Training and reference materials on standards for stakeholder collaboration would help Corps planners achieve more active public participation. This would include meaningful two-way communication and opportunities for direct public inputs into the planning process. Such updated guidance will ensure that Corps projects share common standards for stakeholder involvement/public participation, including expectations that project development processes allow for meaningful public input. **The Corps should conduct a comprehensive review of district-level experiences with stakeholder participation procedures and activities. The Corps should also develop training and reference materials on stakeholder participation standards (Recommendation 8).**

SUMMARY DOCUMENT FOR CORPS PLANNING STUDIES

Corps of Engineers planning studies often include an exceptionally large amount of documentation. It is not unusual for a final feasibility study to be hundreds of pages long and include equally large supplemental documents such as environmental impact statements. This mass of documentation poses many problems: outside parties may find it difficult to locate key assumptions or understand key problems or conclusions; Corps planners themselves may be distracted by computational details and lose sight of "big picture" issues; and external analysts may be challenged to identify key issues they are to provide advice upon.

Inclusion of a summary document in all Corps planning studies would be helpful to interested outside parties, and may also prove beneficial to Corps analysts involved in the study. Such a summary document would allow interested parties from outside the Corps to better and more quickly understand alternatives, assumptions, models, and other important issues. The process of creating this document would also require Corps analysts to articulate all important issues and to express them clearly and succinctly. It would cause Corps analysts to think comprehensively about a planning study and would be valuable in the agency's communication with the public and with interest groups. **A summary document that identifies key environmental and social issues, primary assumptions, alternatives considered and evaluated, objectives sought, benefits and costs (monetized and nonmonetized), trade-offs, and stakeholder perspectives and differences—presented with a consistent format across studies—should be a standard in Corps planning studies (Recommendation 9).**

ENGINEERING METHODS

Three aspects of Corps engineering analysis and methods that relate to changing paradigms of U.S. water resources management, and to changing needs of Corps projects and activities, merit close attention in the years ahead. These are (1) systems engineering aspects of water resource planning, (2) impacts of risk and uncertainty on planning, and (3) integrating engineering methods of analysis with ecosystem restoration planning. Systems engineering focuses on interactions among project components, which may significantly amplify both benefits and costs. The analysis of risk and uncertainty in project planning illuminates the possibly unexpected impacts of deviations from "best estimates" in projections of benefits and costs. The development of engineering methods of analysis for ecosystem restoration provides a way to integrate ecological components within more traditional Corps planning analysis and approaches.

The Corps should strengthen its programs in the areas of systems engineering aspects of water resources, risk and uncertainty analysis, and the integration of engineering and ecosystem analyses.

Part of this strengthening should include the development of updated design manuals that better reflect contemporary methods and theories. These manuals should be used as general guidance rather than as "cookbooks" to specify a series of steps that must be strictly adhered to (Recommendation 10).

INDEPENDENT REVIEW

In today's complex and often politically-charged setting of water resources and environmental management, federal and other agencies vow that management decisions are to be based upon "sound science," but the Corps manages many projects in large, complex ecosystems (e.g., Mississippi River) in which ecological uncertainties can be reduced only so far by additional data or analyses. In addition to environmental uncertainties, many Corps planning studies include forecasts of variables like waterway traffic levels or forecasts of future economic benefits of a given action, such as benefits of flood damages avoided. Such variables are notoriously difficult to accurately forecast, and there are many examples of water projects that fell short of (or in some cases, exceeded) expectations that existed during project promotion and planning.

Given the range of uncertainties and complexities the Corps must cope with in a contemporary planning study, the expertise required to conduct a comprehensive, thorough study may transcend the Corps' abilities and resources. Experts in economics, environmental sciences, and other fields in the nation's universities, research centers, and private sector (and, in some cases, from abroad) could provide useful advice on Corps planning studies. The contemporary reality of U.S. water and natural resources management suggests that some degree of input from external experts is essential in ensuring a degree of "quality control." For example, the 216 study panel on peer review recommended that independent, expert review be conducted in the Corps' more complex and costly planning studies (NRC, 2002).

Not only can independent, expert input be useful in formulating a planning study, advice from external experts can be useful in helping resolve differences of interpretation in post-construction project evaluations. For example, adaptive management principles promote organizational and social learning as part of an iterative process of monitoring outcomes from management and other actions, learning from those outcomes, then using those lessons to adjust future actions. The distillation of clear lessons, however, may founded upon an inability of disparate interest groups to agree upon results of ecosystem monitoring, economics studies, or other relevant information. Input from external experts, from both physical and social sciences, can be useful in clarifying complex results from environmental, economic, and other investigations and in ensuring the progress of adaptive management programs (Jacobs, 2002).

Independent experts can provide fresh perspectives and useful advice on a variety of planning models and approaches, including identification of project alternatives, clarification of key assumptions, economic, engineering, and environmental models, integration of study components and stakeholder perspectives, and evaluation of project impacts and outcomes. Not only can external experts improve Corps planning studies, the use of external experts can stem criticisms of agency self-interest and thereby enhance study credibility. **Independent experts from outside the Corps of Engineers should be enlisted routinely to provide advice in Corps programs and planning studies (Recommendation 11).**

8

Epilogue

There is a long record of formal development of and progress in federal water resources planning methods. The Corps of Engineers was for many years at the forefront of civil engineering practice, and Corps methods and techniques have underpinned the evaluation and development of much of the national infrastructure for flood control and navigation systems. Corps of Engineers planning guidance today is embodied largely within both the federal *Principles and Guidelines* and the Corps *Planning Guidance Notebook.* The *Principles and Guidelines* continues to frame Corps planning methods and the use of data sets, numerical models, and other analytical means. The *P&G,* however, has not been revised since 1983, and some of its analytical and planning precepts date back to the 1960s. This report has emphasized the changed nature of context of Corps projects and planning studies in the late twentieth century. As water resources management concepts and priorities, and social preferences and viewpoints, have broadened, the Corps has been challenged to stay abreast of analytical developments beyond traditional civil engineering concerns. In addition to these changes, the Corps works closely with other federal agencies with important water resources-related responsibilities and there is growing respect for the uncertainties inherent to ecosystems and the difficulties of accurately forecasting the outcomes of human interventions in ecosystems. Finally, the legislative context in which the Corps operates has become increasingly complex.

Against this backdrop—and guided by federal planning guidance in need of updating—the Corps has developed complex agency-specific guidance reflected in the more than 600 page *Planning Guidance Notebook* and scores of economic, engineering, and ecological models. These positive actions, however, have not clarified agency direction or reduced criticisms of the agency. The administration and Congress should rectify inconsistent legislation and set priorities, promote coordination across agencies, and provide leadership in revising federal guidance for the

Corps. The Corps itself, too, must make changes; the agency is experiencing a shift from primarily constructing new civil works projects to (perhaps) primarily operating those civil works projects and is also grappling with possible future roles in ecosystem restoration. Internal adjustments will be necessary to make a successful transition. The Corps may have to undergo changes in organizational culture as well, because its studies and projects are assuming more interdisciplinary dimensions, and there may be additional pressures to better understand the effects of past projects in order to better manage the existing infrastructure. Fundamental legislative changes and a greater attention to inter-agency coordination will be essential. Without this leadership, the Corps will continue to be challenged to resolve conflicts between competing authorizations and competing stakeholder groups, and will continue to lack clarity on which types of planning methods and models the agency should pursue in addressing future water resources needs.

References

AASHTO. 1998. Standard Specifications for Highway Bridges. Washington, D.C.: American Association for Transportation and Highway Officials.

Anderson, R., and P. Kobrin. 2002. Regulatory Economic Analysis at the EPA. Available on-line at *http://yosemite.epa.gov/ee/epalib/ria-epa.nsf;* accessed April 5, 2004.

Apogee Reasearch, Inc. 1996. Monetary Measurement of Environmental Goods and Services: Framework and Summary of Techniques for Corps Planners. IWR Report 96-R-24. Alexandria, VA: Institute for Water Resources.

Arrow, K., M. Cropper, T. Eads, R. Hahn, L. Lave, R. Noll, P. Portney, C. Russell, R. Schmalensee, V. K. Smith, and R. Stavins. 1996. Is there a role for benefit-cost analysis in environmental, health, and safety regulation? Science 272 (1996): 221-222.

Baldwin, W. C. 1999. Engineer Memoirs: Lieutenant General John W. Morris, U.S.A. Retired. EP 870-1-63. Available on-line at *http://www.usace.army.mil/inet/usace-docs/eng-pamphlets/ep870-1-63/entire.pdf,* accessed July 1, 2004.

Bateman, I. J., E. Ozdemirogly, R. Carson, Lee, and S. Mourato. 2002. Economic Valuation with Stated Preference Techniques: A Manual. Brookfield: Elgar Publishing, Inc.

Bjornstad, D. J., and J. R. Kahn. 1996. The Contingent Valuation of Environmental Resources: Methodological Issues and Research Needs. Brookfield: Edward Elgar.

Braden, J. B., and C.D. Kolstad. 1991. Measuring the Demand for Environmental Quality. Amsterdam: North-Holland.

CEN (Comité Européen Normalisation). 1993. Eurocode 7. Comité Européen Normalisation, Technical Committee 250, Brussels.

Champ, P. A., Boyle, K., and T. C. Brown (eds.). 2003. A Primer on Nonmarket Valuation. Boston, MA: Kluwer Academic Press.

Chao, P. T., J. M. Floyd, and W. Holliday. 1997. Empirical Studies of the Effect of Flood Risk on Housing Prices. IWR Report 96-PS-2.

Alexandria, VA: Institute for Water Resources, U.S. Army Corps of Engineers. Review Draft.

Chivers, J., and N. E. Flores. 2002. Market failure in information: The National Flood Insurance Program. Land Economics 78 (4): 515-521.

Cortner, H. J. 1993. Reconciling Citizen, Analyst, and Manager Roles in Democratic Governance: Public Involvement Challenges in the 1990s. Alexandria, VA: Institute for Water Resources, U.S. Army Corps of Engineers.

Creighton, J. L. 1982. Public Involvement Advanced Course Workbook. Ft. Belvoir, VA: Institute for Water Resources, U.S. Army Corps of Engineers.

Delli Priscolli, J. 1978. Why the Federal and Regional Interest in Public Involvement in Water Resources Development. IWR Working Paper 78-1. Alexandria, VA: Institute for Water Resources, U.S. Army Corps of Engineers.

Delli Priscolli, J. 1993. Why the federal and regional interest in public involvement in water resources development. Pp. 61-69 In J. L. Creighton, Jerome Delli Priscoli and C. Mark Dunning (eds.), Public Involvement Techniques: A Reader of Ten Years Experience at the Institute for Water Resources. IWR Research Report 82-R-1. Alexandria, VA: Institute for Water Resources, U.S. Army Corps of Engineers.

Eckstein, O. 1958. Water Resource Development: The Economics of Project Evaluation. Cambridge, MA: Harvard University Press.

EPA (Environmental Protection Agency). 2000. Guidelines for Preparing Economic Analyses. Washington, D.C.: Environmental Protection Agency.

ESMAP (Joint UNDP/World Bank Energy Sector Management Assistance Programme). 2003. Stakeholder Involvement in Options Assessment: Promoting Dialogue in Meeting Water and Energy Needs. Washington, D.C.: The World Bank Group.

Executive Order 12,291. 1981. Regulatory Impact Analysis and Review. Washington, D.C.: White House.

Executive Order 12,866. 1993. Regulatory Planning and Review. Washington, D.C.: White House.

Fastabend, D. 2002. Memorandum to General Flowers regarding opening statement to the House/Senate Committee Hearing, Washington, D.C.

FEMA (Federal Emergency Management Agency). 2003. Benefit-Cost Riverine Full Data Module User's Guide. Federal Insurance Mitigation Administration. Available on-line at *http://www .fema.gov/fima/ hmgp/riverine_a.shtm;* accessed March 2003.

Freeman, A. M. 1994. The Measurement of Environmental and Resource Values: Theory and Methods. Washington, D.C.: Resources for the Future.

Garrod, G., and K.G. Willis. 1999. Economic Valuation of the Environment: Methods and Case Studies. Cheltenham, UK: Edward Elgar.

Griffin, R. 2002. E-mail Message on Planning Capability from Major General Griffin, Director of Civil Works, to Division Commanders, 7 July.

Gore, J. A., and F. D. Shields. 1995. Can large rivers be restored? A focus on rehabilitation. Bioscience 45: 142-152.

Haab, T. C., and K. E. McConnell. 2002. Valuing Environmental and Natural Resources: The Econometrics of Non-market Valuation. Brookfield: Edward Elgar.

Hanchey, J. R. 1988. A decade of contribution. Pp. 7-11 In Creighton, J. L., J. Delli Priscoli, and C. M. Dunning (eds.) Public Involvement Techniques: A Reader of Ten Years Experience at the Institute for Water Resources. IWR Research Report 82-R-1. Alexandria, VA: Institute for Water Resources, U.S. Army Corps of Engineers.

Hanemann, W. M. 1994. Valuing the Environment through Contingent Valuation. Journal of Economic Perspectives 8(4): 19-43.

Hanley, N., and C. L. Spash. 1993. Cost-Benefit Analysis and the Environment. Cheltenham, UK: Edward Elgar.

Holmes, B. H. 1972. A History of Federal Water Programs, 1800-1960. Miscellaneous Publication 1233. Washington, D.C.: U.S. Department of Agriculture.

Holoway, J. M., and R. Burby. 1990. The effects of floodplain development controls on residential land values. Land Economics 66 (3): 259-270.

IUCN. 2000. Vision for Water and Nature: A World Strategy for Conservation and Sustainable Management of Water Resources. Cambridge, MA: IUCN Publications.

IWR (Institute for Water Resources). 1996. Evaluation of Environmental Investment Procedures: Overview Manual. IWR Report 96-R-30. Alexandria, VA: Institute for Water Resources, U.S. Army Corps of Engineers.

IWR. 1997. Evaluation of Environmental Investment Program. Revised draft document. November.

Jacobs, J. W. 2002. Broadening U.S. water resources project planning and evaluation. Natural Resources Journal 42(1): 21-31.

Klosterman, M. J., and J. Sanders. Application of Risk-Based Analysis to Seepage Modification Walter F. George Lock & Dam, Georgia. 18th USCOLD Annual Meeting, NY.

Kneese, A. B. 1993. Economics and water resources. Pp. 23-35 In Reuss, M. (ed.) Water Resources Administration in the United States: Policy, Practices, and Emerging Issues. East Lansing, MI: American Water Resources Association and Michigan State University Press.

Kopp, R. J., W. W. Pommerehne, and N. Schwarz. Determining the Value of Non-marketed Goods: Economic, Psychological, and Policy Relevant Aspects of Contingent Valuation Methods. Boston, MA: Kluwer Academic Press.

Krutilla, J. V., and O. Eckstein. 1958. Multiple Purpose River Development: Studies in Applied Economic Analysis. Baltimore, MD: Johns Hopkins.

Langton, S. 1993. Historical profile of public involvement in the U.S. Army Corps of Engineers: Key milestones since 1960. Pp. 4-31 In Cerighton, J. L., C. M. Dunning, J. Delli Priscoli, and D. B. Ayres (eds.) Public Involvement and Dispute Resolution: A Reader on the Second Decade of Experience at the Institute for Water Resources. IWR Report 98-R-5. Alexandria, VA: Institute for Water Resources, U.S. Army Corps of Engineers.

Lind, R. C., K. J. Arrow, G. R. Correy, P. Dasgupta, A. K. Sen, T. Stauffer, J. E. Stiglitz, J. A. Stockfisch, and R. Wilson. 1982. Discounting for Time and Risk in Energy Policy. Washington, D.C.: Resources for the Future.

Loomis, J. B. 1997. Use of non-market valuation methods in water resources management assessments. Water Resources Update 109 (Autumn): 5-9.

Maass, A., M. M. Hufschmidt, R. Dorfman, H. A. Thomas, Jr., S. A. Marglin, and G. M. Fair. 1962. Design of Water-Resource Systems: New Techniques for Relating Economic Objectives, Engineering Analysis, and Governmental Planning. Cambridge, MA: Harvard University Press.

Martin, L., and E. Stakhiv. 1999. Sustainable Development: Concepts, Goals, and Relevance to the Civil Works Program. IWR Report 99-PS-1. Alexandria, VA: Institute of Water Resources, U.S. Army Corps of Engineers.

Mazmanian, D. A., and J. Nienaber. 1979. Can Organizations Change?

Environmental Protection, Citizen Participation, and the Corps of Engineers. Washington, D.C.: The Brookings Institute.

Middleton, B. 1999. Wetland restoration, flood pulsing, and disturbance dynamics. New York, NY: John Wiley & Sons.

Mol, A. P. J. 2001. Globalization and Environmental Reform: The Ecological Modernization of the Global Economy. Cambridge, MA: The MIT Press.

NCPS (National Commission on the Public Service). 2003. Urgent Business for America. Revitalizing the Federal Government for the 21st Century. Washington, D.C.: Brookings Institute for Public Service.

NRC (National Research Council). 1992. Restoration of aquatic ecosystems: science, technology and public policy. Washington, D.C.: National Academy Press.

NRC. 1995. Flood Risk Management and the American River Basin: An Evaluation. Washington, D.C.: The National Academies Press.

NRC. 1996. Understanding Risk: Informing Decision in a Democratic Society. Washington, D.C.: The National Academies Press.

NRC. 1999a. New Directions in Water Resources Planning for the U.S. Army Corps of Engineers. Washington, D.C.: National Academy Press.

NRC. 1999b. New Strategies for America's Watersheds. Washington, D.C.: National Academy Press.

NRC. 2000. Risk Analysis and Uncertainty in Flood Damage Reduction studies. Washington, D.C.: The National Academies Press.

NRC. 2001. Inland Navigation System Planning: The Upper Mississippi River—Illinois Waterway. Washington, D.C.: National Academy Press.

NRC. 2002a. The Missouri River Ecosystem: Exploring the Prospects for Recovery. Washington, D.C.: National Academy Press.

NRC. 2002b. Review Procedures for Water Resource Planning. Washington, D.C.: The National Academies Press.

NWC (National Water Commission). 1973. New directions in U.S. Water Policy: Summary Conclusions and Recommendations at 45. Washington, D.C.: U.S. Government Printing Office.

OECD. 1973. Water Management and the Environment. Paris: OECD.

OMB (Office of Management and Budget). 2000. Guidelines to Standardize Measures of Costs and Benefits and the Format of Accounting Statements. Washington, D.C.: OMB. Available on-line at *http://www.whitehouse.gov/omb/memoranda/m00-08.pdf;* accessed

April 5, 2004.

OMB. 2003. Draft 2003 Report to Congress on the Costs and Benefits of Federal Regulations. Federal Register 68(22, February 3): 5492-5527.

Popovic, N. A. F. 1993. The right to participate in decisions that affect the environment. Pace Environmental Law Review 10(2): 683-709.

Porter, T. M. 1996. Trust in Numbers: The Pursuit of Objectivity in Science and Public Life. Princeton, NJ: Princeton University Press.

Randall, A. 1999. Making the Environment Count. Brookfield: Edward Elgar.

Reike, B., and D. Kenney. 1997. Resource Management at the Watershed Level. Boulder, CO: Natural Resources Law Center.

Rhoads, B. L., and E. E. Herricks. 1996. Naturalization of headwater agricultural streams in Illinois: Challenges and possibilities. Pp. 331-367 In Brookes, A., and D. Shields (eds.) River Channel Restoration. Chichester: Wiley.

Rosenbaum, W. A. 1979. Public participation: Required, but is it important? Citizen Participation 1(1).

Rosenbaum, W. A. 1991. Environmental Politics and Policy. Washington, D.C.: Congressional Quarterly Press.

SER (Society for Ecological Restoration). 2002. The SER primer on ecological restoration. Society for Ecological Restoration Science and Policy Working Group. Available on-line at: *http://www.ser.org*.

Shabman, L. 1993. Environmental Activities in Corps of Engineers Water Resources Programs: Charting a New Direction. IWR Report 93-PS-1. Alexandria, VA: Institute of Water Resources.

Shabman, L., A. Riley, and G. Stedge. 1997. Evaluation of floodplain permanent evacuation measures: An alternative approach for the U.S. Army Corps of Engineers. Report prepared for the U.S. Army Corps of Engineers. Fort Belvoir, VA: Institute for Water Resources.

Shallat, T. 2000. Structures in the Stream: Water, Science, & the Rise of the U.S. Army Corps of Engineers. Austin: University of Texas Press.

Shallat, T. 2002. Successes through failure: Army science in harbor construction 1820-1860. Water Resources Impact 5(1): 5-7.

Smith, V. K. 1996. Estimating Economic Values for Nature. Brookfield: Edward Elgar.

Stanford, J. A., J. V. Ward, W. J. Liss, C. A. Frissell, R. N. Williams, J. A. Lichatowich and C. C. Coutant. 1996. A general protocol for restoration of regulated rivers. Regulated Rivers: Research and Management 12: 391-413.

Stakhiv, E., R. Cole, P. Scodari, and L. Martin. 2003. Improving environmental benefits analysis in ecosystem restoration planning. IWR Report 03-PS-3. Alexandria, VA: Institute for Water Resources.

USACE (U.S. Army of Corps of Engineers). 1979. Missouri River Main Stem Reservoir System Reservoir Regulation Manual (Master Manual). Omaha, NE: U.S. Army Engineer Division, Missouri River Corps of Engineers.

USACE. 1981. Feasibility Report, Mississippi River at St. Paul Minnesota, Reevaluation of St. Paul Flood Control Project. St. Paul, MN: U.S. Army Corps of Engineers.

USACE . 1992. Guidelines for risk and uncertainty analysis in water resources planning. 2 Volumes. IWR-92-R-1. Alexandria, VA: Institute for Water Resources, U.S. Army Corps of Engineers.

USACE. 1998. Civil Works Overview. Available on-line at *http://www.iwr.usace.army.mil/iwr/plannersweb/docs/24;* accessed February 11, 2004..

USACE. 1999a. Economic Restoration—Supporting Policy Information. EP 1165-2-502. Washington, D.C.: U.S. Army Corps of Engineers.

USACE. 1999b. Nonstructural Flood Damage Reduction Projects. Available on-line at *http://www.usace.army.mil/inet/functions/cw/cecwp/branches/mp_and_dev/nonstr.pdf;* accessed February 11, 2004.

USACE. 2000. Planning Guidance Notebook. ER 1105-2-100, April 22, 2000. Available on-line at *http://www.usace.army.mil/inet/usace-docs/eng-regs/er1105-2-100/toc.htm;* accessed February 11, 2004.

USACE. 2001a. Civil Works Programs. IWR Report 98-R-3. Fort Belvoir, VA: U.S. Army Corps of Engineers, Institute for Water Resources, U.S. Army Corps of Engineers.

USACE. 2001b. Great Egg Harbor Inlet to Townsends Inlet, Draft Feasibility Report and Integrated Environmental Impact Statement. CENAP-PL-E-01-04. Available by calling (215) 656-6555, Philadelphia District Office.

USACE. 2001c. Non-Structural Flood Damage Reduction Within the Corps of Engineers: What Districts Are Doing. Washington, D.C.: USACE National Nonstructural/Floodproofing Committee.

USACE. 2001d. Project Partnership Kit. IWR Report No. 96-R-10, January. Available on-line at *http://www.iwr.usace.army.mil/plannersweb/docs/ppkit.pdf*; accessed February 11, 2004.

February 11, 2004.

USACE. 2002a. Economic Guidance Memorandum 02-04: Unit Day Values for Recreation, Fiscal Year 2002. Available on-line at *http://www.usace.army.mil/inet/functions/cw/cecwp/General_guidance/egm02-04.pdf*; accessed February 11, 2004.

USACE. 2002b. Upper Mississippi river—Illinois waterway system navigation study Newsletter. St. Paul, MN. Volume 8(2) August 2002.

USACE. 2002c. Summary of Water Project Cost Sharing: Non-Federal Responsibilities. Civil Works Program Development and Execution Course. Washington, D.C.: U.S. Army Corps of Engineers.

USACE. 2003. Authorized Uncompleted Projects. January 9, 2003. Washington, D.C.: U.S. Army Corps of Engineers.

WRC (U.S. Water Resources Council). 1973. Water and Related Land Resources: Establishment of Principles and Standards for Planning. Federal Register 38: 24784,248222-248223.

WRC. 1983. Economic and Environmental Principles and Guidelines for Water and Related Land Resources Implementation Studies. Washington, D.C.: U.S. Government Printing Office.

White, G. F. 1971. Unpublished paper, delivered at Columbia University, 21 March.

White, G. F. 1945. Human Adjustment to Floods. University of Chicago Department of Geography Research Papers No. 29.

World Bank, Operations Policy Department. 1996. Handbook on Economic Analysis of Investment Projects. Washington, D.C.: World Bank.

World Bank. 1999. Operational Manual. Washington, D.C.: World Bank.

World Bank and International Finance Corporation. 2000. Chad-Cameroon Petroleum Development and Pipeline. Appraisal document 19343, Washington, D.C. Available on-line at *http://www.worldbank.org/afr/ cproj/project/tdpppadann3.pdf*; accessed February 11, 2004.

World Commission on Dams. 2000. Dams and Development: A New Framework for Discussion. London: Earthscan Publications.

Zerbe, R. O., Jr., and D. D. Dively. 1994. Benefit Cost Analysis in Theory and Practice. New York: Harper Collins.

Appendixes

Appendix A

Water Resources Development Act 2000 Public Law No. 106-541, of the 106th Congress

SEC. 216. NATIONAL ACADEMY OF SCIENCES STUDY.

(a) DEFINITIONS—In this section, the following definitions apply:

(1) ACADEMY—The term "Academy" means the National Academy of Sciences.

(2) METHOD—The term "method" means a method, model, assumption, or other pertinent planning tool used in conducting an economic or environmental analysis of a water resources project, including the formulation of a feasibility report.

(3) FEASIBILITY REPORT—The term "feasibility report" means each feasibility report, and each associated environmental impact statement and mitigation plan, prepared by the Corps of Engineers for a water resources project.

(4) WATER RESOURCES PROJECT—The term "water resources project"' means a project for navigation, a project for flood control, a project for hurricane and storm damage reduction, a project for emergency streambank and shore protection, a project for ecosystem restoration and protection, and a water resources project of any other type carried out by the Corps of Engineers.

(b) INDEPENDENT PEER REVIEW OF PROJECTS—

(1) IN GENERAL—Not later than 90 days after the date of enactment of this Act, the Secretary shall contract with the Academy to study, and make recommendations relating to, the independent peer review of feasibility reports.

(2) STUDY ELEMENTS—In carrying out a contract under paragraph (1), the Academy shall study the practicality and efficacy of the independent peer review of the feasibility reports, including—

(A) the cost, time requirements, and other considerations relating to the implementation of independent peer review; and

(B) objective criteria that may be used to determine the most effective application of independent peer review to feasibility reports for each type of water resources project.

(3) ACADEMY REPORT—Not later than 1 year after the date of a contract under paragraph (1), the Academy shall submit to the Secretary, the Committee on Transportation and Infrastructure of the House of Representatives, and the Committee on Environment and Public Works of the Senate a report that includes—

(A) the results of the study conducted under paragraphs (1) and (2); and

(B) in light of the results of the study, specific recommendations, if any, on a program for implementing independent peer review of feasibility reports.

(4) AUTHORIZATION OF APPROPRIATIONS—There is authorized to be appropriated to carry out this subsection $1,000,000, to remain available until expended.

(c) INDEPENDENT PEER REVIEW OF METHODS FOR PROJECT ANALYSIS—

(1) IN GENERAL—Not later than 90 days after the date of enactment of this Act, the Secretary shall contract with the Academy to conduct a study that includes—

(A) a review of state-of-the-art methods;

(B) a review of the methods currently used by the Secretary;

(C) a review of a sample of instances in which the Secretary has applied the methods identified under subparagraph (B) in the analysis of each type of water resources project; and

(D) a comparative evaluation of the basis and validity of state-of-the-art methods identified under subparagraph (A) and the methods identified under subparagraphs (B) and (C).

(2) ACADEMY REPORT—Not later than 1 year after the date of a contract under paragraph (1), the Academy shall transmit to the Secretary, the Committee on Transportation and Infrastructure of the House of Representatives, and the Committee on Environment and Public Works of the Senate a report that includes—

(A) the results of the study conducted under paragraph (1); and

(B) in light of the results of the study, specific recommendations for modifying any of the methods currently used by the Secretary for conducting economic and environmental analyses of water resources projects.

(3) AUTHORIZATION OF APPROPRIATIONS—There is authorized to be appropriated to carry out this subsection $2,000,000. Such sums shall remain available until expended.

Appendix B

Corps Divisions and Districts

Great Lakes and Ohio River Division (CELRD)

- Buffalo District (CELRB)
- Chicago District (CELRC)
- Detroit District (CELRE)
- Huntington District (CELRH)
- Louisville District (CELRL)
- Nashville District (CELRN)
- Pittsburgh District (CELRP)

Mississippi Valley Division (CEMVD)

- Memphis District (CEMVM)
- New Orleans District (CEMVN)
- Rock Island District (CEMVR)
- St. Louis District (CEMVS)
- St. Paul District (CEMVP)
- Vicksburg District (CEMVK)

North Atlantic Division (CENAD)

- Baltimore District (CENAB)
- Europe District (CENAU)
- New England District (CENAE)
- New York District (CENAN)
- Norfolk District (CENAO)
- Philadelphia District (CENAP)

Northwestern Division (CENWD)

- Kansas City District (CENWK)
- Omaha District (CENWO)
- Portland District (CENWP)
- Seattle District (CENWS)
- Walla Walla District (CENWW)

Pacific Ocean Division (CEPOD)

- Alaska District (CEPOA)
- Far East District (CEPOF)
- Honolulu District (CEPOH)
- Japan Engineer District (CEPOJ)

South Atlantic Division (CESAD)

- Charleston District (CESAC)
- Jacksonville District (CESAJ)
- Mobile District (CESAM)
- Savannah District (CESAS)
- Wilmington District (CESAW)

South Pacific Division (CESPD)

- Albuquerque District (CESPA)
- Los Angeles District (CESPL)
- Sacramento District (CESPK)
- San Francisco District (CESPN)

Southwestern Division (CESWD)

- Fort Worth District (CESWF)
- Galveston District (CESWG)
- Little Rock District (CESWL)
- Tulsa District (CESWT)

Appendix C

Office of Management and Budget's Guidance on Nonmarket Valuation Techniques

How should I value benefits that are indirectly traded in markets? Some benefits correspond to goods or services that are indirectly traded in the marketplace. Their value is reflected in the prices of related goods that are directly traded. Examples include reductions in health-and-safety risks, the use-values of environmental amenities (for example, recreational fishing or hiking and camping), and the value of improved scenic visibility. You should use willingness-to-pay measures as the basis for estimating the monetary value of such indirectly traded goods. When practical obstacles prevent the use of directly "revealed preference" methods based on actual market behavior to measure willingness-to-pay, you may consider the use of alternative "stated preference" methods based on survey techniques.

How should I value goods that are not traded directly or indirectly in markets? Some types of goods—such as preserving environmental or cultural amenities apart from their use and direct enjoyment by people (their so-called "nonuse" value)—are not traded directly or indirectly in markets. Estimation of the benefits for these types of goods is even more difficult than for indirectly traded goods, because market-related transactions do not exist to provide data for willingness-to-pay estimates.

Stated preference methods using survey techniques, such as contingent valuation methods, may provide the only analytical approach currently available for estimating the values of many of these goods, particularly goods providing "nonuse" values. The lack of observable behavior for these goods, combined with their complex and often unfamiliar nature, calls for careful design and execution of these surveys. Confidence in their results requires rigorous analysis of the responses and full characterization of uncertainties. The use of studies that rely on the state of the art in survey design and implementation is important to assuring confidence in the results. In addition, these studies should satisfy checks on their internal consistency. For example, you should apply a "scope" test to show that individuals are willing to pay more for incrementally greater amounts of a good.

SOURCE: Office of Management and Budget (OMB, 2000, pp.10-11).

Does the proposal permit or encourage use of methods of "contingent valuation" when quantifying the benefits and costs of environmental rules? Answer: Yes. The proposal recognizes that in certain situations important benefits arising from regulation affect goods or attributes that are not directly traded in markets (e.g., the improvement of “visibility” or the preservation of pristine wilderness areas). In these cases, the contingent valuation approach may provide the most suitable methodology for estimating such values. Where the analyst elects to use contingent valuation methods, the proposal cautions the analyst to follow several "best practices" to ensure that the resulting estimated values are reasonable and valid.

Does the proposal permit agencies to present qualitative information about benefits and costs in situations where quantification is not feasible? Answer: Yes. OMB has always encouraged the agencies to do so. This draft continues that practice.

SOURCE: OMB (2003).

Appendix D

Analysis of Nonstructural Flood Damage Reduction

A paper by Shabman et al. (1997) provides an illustration of the peculiar economic position of nonstructural flood mitigation. A parcel with structures is subject to flooding. The choice is whether to build a levee or purchase the property for evacuation. Here is their example (all monetary values in thousands) in simplified form:

Market value of land without flooding: L = \$9
Market value of structure (not flooded): S = \$2
Present value of expected flood damage to land: D = \$3
Cost of levee: C = \$2
Benefits of evacuated land: E = \$4.5

Net benefits of levee (damages mitigated less cost of mitigation): $D - C$ = \$1.

Net benefits of evacuation (benefits of evacuated land less cost of land and structure adjusted for expected damage): $E - (L + S - D)$ = -\$3.5.

Evacuation looks unfavorable because it removes \$8 in developed property value (after capitalizing in the expected damages) and returns only \$4.5 in undeveloped benefits. The fundamental issue is that today's assessment is the result of the past. Consider the following ways in which historical values might deviate from future values:

(1) The value of the land and structure is reflective of various types of policies such as access to local utilities and services, agricultural subsidies, and flood protection or compensation subsidies.

(a) Say that, without crop subsidies, the flood damages would be cut in half, to D' = \$1.5. In that case, $D' - C$ = -\$0.5; the levee would not make sense either.

(b) Without crop subsidies, the value of the land and structures might also decrease. Say the devaluation is to $L' = \$5$ and $S' = \$1$, respectively. Then, $E - (L' + S' - D') = \$0$. In other words, history and other policies matter.

(2) Consider a case in which the floodplain land has not yet been cleared and a structure built. Say that clearing the land and building the structure would cost \$6 (call it R), producing net returns $L + S - D - R =$ \$2 (i.e., assuming flooding remains). Meanwhile, doing so would eliminate \$4.5 in benefits from the undeveloped land. In this case, the private market would leave the land undeveloped unless the flood protection project subsidizes its development

The Corps' methods are supposed to be future-oriented and to adjust for value distortions due to policies or other factors. This is often very difficult to do.

Reference

Shabman, L., A. Riley, and G. Stedge. 1997. Evaluation of floodplain permanent evacuation measures: An alternative approach for the U.S. Army Corps of Engineers. Report prepared for the U.S. Army Corps of Engineers. Fort Belvoir, VA: Institute for Water Resources.

Appendix E

Army Corps of Engineers Planning Center of Expertise

CECW-P (1105-10b)

MEMORANDUM FOR COMMANDERS, MAJOR SUBORDINATE COMMANDS

SUBJECT: Planning Centers of Expertise

1. Reference memorandum dated 16 April 2003, subject: Planning Centers of Expertise.

2. The referenced memorandum stated our intent to name USACE Planning Centers of Expertise and requested information to make such a determination. Your response have been reviewed and the following designation has been decided for the five key business functions:

Inland Navigation—CELRD
Deep Draft Navigation—CESAD
Flood Damage Reduction—CESPD
Hurricane and Storm Damage Prevention—CENAD
Ecosystem Restoration—CEMVD

3. I have also chosen CESWD to be the USACE Planning Center of Expertise for Water Supply and Reallocation and I am acknowledging CENWD as the National Hydropower Planning Center of Expertise.

4. In addition, the U.S. Army Corps of Engineers National Nonstructural/Flood Proofing Committee (NFPC) is an excellent support element for Nonstructural Flood Damage Reduction. The NFPC has the capability to provide assistance to truly innovative planning, including nonstructural flood damage reduction, flood plain management, ecosystem restoration, and combined NED/NER planning. The NFPC can provide sup-

port for at least three of the five key business functions: flood damage reduction, hurricane and storm damage prevention, and ecosystem restoration.

5. Enclosed for your information and use is a roles statement for USACE Planning Centers of Expertise.

FOR THE COMMANDER:

Encl

ROBERT H. GRIFFIN
Major General, USA
Acting Director of Civil Works

DISTRIBUTIONS:
(See Page 2)
CECW-P (1105-2-10b)
SUBJECT: Planning Centers of Expertise

DISTRIBUTIONS:
Commander, Great Lakes and Ohio River Division, CELRD
Commander, Mississippi Valley Division, CEMVD
Commander, North Atlantic Division, CENAD
Commander, Northwestern Division, CENWD
Commander, Pacific Ocean Division, CEPOD
Commander, South Atlantic Division, CESAD
Commander, South Pacific Division, CESPD
Commander, Southwestern Division, CESWD

PLANNING CENTER OF EXPERTISE ROLE*

An USACE planning center of expertise will primarily focus on plan formulation and the complex technical evaluation associated with the plan formulation. The center will provide support in the following areas:

- The center will provide consulting services and at the direction of HQUSACE the center would accomplish very costly, highly complex and controversial studies or key analytical components of very costly, highly complex and controversial studies.
- The center would provide accomplishment of key analytical components of studies as directed by the Planning Chiefs Advisory Board for multi-region/national efforts.
- The center would provide independent review support, to supplement the capabilities of the MSC regional planning expertise centers. This would be especially important in those cases where an MSC can not satisfy the inter-district review requirement when the expertise within the MSC is limited to a single physical location.
- The center would provide advice to HQUSACE, the laboratories and other stakeholders on significant regional and nation-wide planning issues.
- The center would assist in establishing research and development priorities in the mission area, coordinating the recommendations of the Planning Chiefs Advisory Board and with the established MSC lead that coordinates the review of research and development initiatives for the mission area.
- The center would be a proponent for training opportunities related to the assigned mission area.
- The center would manage a program of sharing lessons through coordination with the MSC regional planning expertise centers, sponsoring workshops and technology transfer.
- The center would supplement the HQUSACE staff in policy compliance review on projects where the center has had no prior participation, on an as requested reimbursable basis at times of very high workload such as before a potential WRDA.

* The Army Corps of Engineers has established the USACE Technical Excellence Network, a gateway (http://ten.usace.army.mil/techexnet.aspx) to share information and experience on topics related to the mission of the Corps.

- The center would enhance basic planning expertise throughout the Corps by providing shadowing opportunities of individuals with specialized planning expertise.

Appendix F

Rosters

Coordinating Committee

LEONARD SHABMAN, Chair, Resources for the Future, Washington, D.C.
GREGORY B. BAECHER, University of Maryland, College Park
DONALD F. BOESCH, University of Maryland, Cambridge
GERALDINE KNATZ, Port of Long Beach, Long Beach, California
JAMES K. MITCHELL, Virginia Polytechnic Institute and State University, Blacksburg
A. DAN TARLOCK, Chicago-Kent College of Law, Chicago, Illinois
VICTORIA J. TSCHINKEL, The Nature Conservancy, Altamonte Springs, Florida
JAMES G. WENZEL, Marine Development Associates, Inc., Saratoga, California
M. GORDON WOLMAN, Johns Hopkins University, Baltimore, Maryland

Peer Review Procedures

JAMES K. MITCHELL, Chair, Virginia Polytechnic Institute and State University, Blacksburg
MELBOURNE BRISCOE, Office of Naval Research, Arlington, Virginia
STEPHEN J. BURGES, University of Washington, Seattle
LINDA CAPUANO, Honeywell, Inc., San Jose, California
DENISE FORT, University of New Mexico, Albuquerque
PORTER HOAGLAND, Woods Hole Oceanographic Institution, Massachusetts
DAVID H. MOREAU, University of North Carolina, Chapel Hill
CRAIG PHILIP, Ingram Barge Company, Nashville, Tennessee
JOHN T. RHETT, Consultant, Arlington, Virginia
RICHARD E. SPARKS, Illinois Water Resources Center, Urbana
BORY STEINBERG, Steinberg and Associates, McLean, Virginia

Panel on Methods and Techniques of Project Analysis

GREGORY B. BAECHER, Chair, University of Maryland, College Park
JOHN B. BRADEN, University of Illinois, Urbana-Champaign
DAVID L. GALAT, University of Missouri, Columbia
GERALD E. GALLOWAY, Titan Corporation, Fairfax, Virginia
ROBERT G. HEALY, Duke University, Durham, North Carolina
EDWIN E. HERRICKS, University of Illinois, Urbana-Champaign
CATHERINE L. KLING, Iowa State University, Ames
LINDA A. MALONE, College of William and Mary, Williamsburg, Virginia
RAM MOHAN, Blasland, Bouck & Lee, Inc., Annapolis, Maryland
MAX J. PFEFFER, Cornell University, Ithaca, New York
DOUG PLASENCIA, AMEC, Phoenix, Arizona
DENISE J. REED, University of New Orleans, Louisiana
JAN A. VELTROP, Consultant, Skokie, Illinois

Adaptive Management for Resource Stewardship

DONALD F. BOESCH, Chair, University of Maryland-Center for Environmental Science, Cambridge
HENRY J. BOKUNIEWICZ, University of New York, Stony Brook
RICHARD DE NEUFVILLE, Massachusetts Institute of Technology, Cambridge
G. EDWARD DICKEY, Consultant, Baltimore, Maryland
HOLLY D. DOREMUS, University of California, Davis
CARL H. HERSHNER, Virginia Institute of Marine Science, Gloucester Point, Virginia
FREDRICK J. HITZHUSEN, Ohio State University, Columbus
CHARLES D. D. HOWARD, Charles Howard Associates, British Columbia, Canada
BARRY R. NOON, Colorado State University, Fort Collins
THAYER SCUDDER, California Technology Institute, Pasadena
ROBERT W. STERNER, University of Minnesota, Minneapolis

River Basin and Coastal Systems Planning

PETER R. WILCOCK, Chair, Johns Hopkins University, Baltimore, Maryland
GAIL M. ASHLEY, Rutgers University, Piscataway, New Jersey
DENISE L. BREITBURG, Smithsonian Environmental Research Center, Edgewater, Maryland
VIRGINIA R. BURKETT, U.S. Geological Survey, Lafayette, Louisiana

JOSEPH J. CORDES, George Washington University, Washington, D.C.
ROBERT G. DEAN, University of Florida, Gainesville
JOHN A. DRACUP, University of California, Berkeley
WILLIAM J. MITSCH, Ohio State University, Columbus
ROBERT E. RANDALL, Texas A&M University, College Station
A. DAN TARLOCK, Chicago Kent College of Law, Chicago, Illinois

Water Science and Technology Board
RICHARD G. LUTHY, Chair, Stanford University, Stanford, California
JOAN B. ROSE, Vice Chair, Michigan State University, East Lansing
RICHELLE M. ALLEN-KING, State University of New York (SUNY), Buffalo, New York
GREGORY B. BAECHER, University of Maryland, College Park
KENNETH R. BRADBURY, Wisconsin Geological and Natural History Survey, Madison
JAMES CROOK, Water Reuse Consultant, Norwell, Massachusetts
EFI FOUFOULA-GEORGIOU, University of Minnesota, Minneapolis
PETER GLEICK, Pacific Institute for Studies in Development, Environment, and Security, Oakland, California
JOHN LETEY, JR., University of California, Riverside
DIANE M. MCKNIGHT, University of Colorado, Boulder (*through June 30, 2003*)
CHRISTINE L. MOE, Emory University, Atlanta, Georgia
ROBERT PERCIASEPE, National Audubon Society, Washington, D.C.
RUTHERFORD H. PLATT, University of Massachusetts, Amherst (*through June 30, 2003*)
JERALD L. SCHNOOR, University of Iowa, Iowa City
LEONARD SHABMAN, Resources for the Future, Washington, D.C.
R. RHODES TRUSSELL, Trussell Technologies, Inc., Pasadena, California
KARL K. TUREKIAN, Yale University, New Haven, Connecticut
HAME M. WATT, Independent Consultant, Washington, D.C.
JAMES L. WESCOAT, JR., University of Illinois at Urbana-Champaign

Water Science and Technology Board Staff
STEPHEN D. PARKER, Director
LAURA J. EHLERS, Senior Staff Officer
JEFFREY W. JACOBS, Senior Staff Officer
WILLIAM S. LOGAN, Senior Staff Officer
LAUREN E. ALEXANDER, Staff Officer

MARK C. GIBSON, Staff Officer
STEPHANIE E. JOHNSON, Staff Officer
M. JEANNE AQUILINO, Administrative Associate
ELLEN A. DE GUZMAN, Research Associate
PATRICIA JONES KERSHAW, Study/Research Associate
ANITA A. HALL, Administrative Assistant
JON Q. SANDERS, Senior Project Assistant
DOROTHY K. WEIR, Project Assistant

Ocean Studies Board

NANCY RABALAIS (*Chair*), Louisiana Universities Marine Consortium, Chauvin
LEE G. ANDERSON, University of Delaware, Newark
WHITLOW AU, University of Hawaii at Manoa
ARTHUR BAGGEROER, Massachusetts Institute of Technology, Cambridge
RICHARD B. DERISO, Inter-American Tropical Tuna Commission, La Jolla, California
ROBERT B. DITTON, Texas A&M University, College Station
EARL DOYLE, Shell Oil (Retired), Sugar Land, Texas
ROBERT DUCE, Texas A&M University, College Station
PAUL G. GAFFNEY, II, National Defense University, Washington, D.C.
WAYNE R. GEYER, Woods Hole Oceanographic Institution, Massachusetts
STANLEY R. HART, Woods Hole Oceanographic Institution, Massachusetts
RALPH S. LEWIS, Connecticut Geological Survey (Retired), Hartford
WILLIAM F. MARCUSON, III, U.S. Army Corp of Engineers (Retired), Vicksburg, Mississippi
JULIAN P. MCCREARY, JR., University of Hawaii, Honolulu
JACQUELINE MICHEL, Research Planning, Inc., Columbia, South Carolina
JOAN OLTMAN-SHAY, Northwest Research Associates, Inc., Bellevue, Washington
ROBERT T. PAINE, University of Washington, Seattle
SHIRLEY A. POMPONI, Harbor Branch Oceanographic Institution, Fort Pierce, Florida
FRED N. SPIESS, Scripps Institution of Oceanography, La Jolla, California
DANIEL SUMAN, Rosentiel School of Marine and Atmospheric Science, Unviersity of Miami, Florida

Ocean Studies Board Staff
SUSAN ROBERTS, Director
JENNIFER MERRILL, Senior Program Officer
DAN WALKER, Senior Program Director
JOANNE BINTZ, Program Officer
ALAN B. SIELEN, Visiting Scholar
ANDREAS SOHRE, Financial Associate
SHIREL SMITH, Administrative Associate
JODI BACHIM, Senior Project Assistant
NANCY CAPUTO, Senior Project Assistant
SARAH CAPOTE, Project Assistant

Appendix G

Biographical Information of Panel Members and Staff

Gregory B. Baecher, *Chair,* is a professor in the civil engineering program at the University of Maryland. Prior to this, Dr. Baecher served on the faculty of civil engineering at the Massachusetts Institute of Technology from 1976 to 1988, and he served as the CEO and founder of Con-Solve Incorporated, Lexington, Massachusetts, from 1988 to 1995. His fields of expertise include risk analysis, water resources engineering, and statistical methods. Dr. Baecher is currently a member of the Water Science and Technology Board. Dr. Baecher received his B.S. degree in civil engineering from the University of California-Berkeley and his M.S. and his Ph.D. degrees in civil engineering from the Massachusetts Institute of Technology.

John B. Braden is director of the Environment Council and a professor of agricultural and consumer economics at the University of Illinois Urbana-Champaign. He also served as director of the Illinois Water Resources Center and associate provost among other appointments at the University of Illinois. His current research includes economics of river systems, valuation of environmental quality, environment and development. He received a B.A. in economics from Miami University, Ohio, and an M.S., an M.A., and a Ph.D. degrees in agricultural and applied economics from the University of Wisconsin-Madison.

David L. Galat is an associate professor at the Department of Fishery and Wildlife Sciences at the University of Missouri and assistant unit leader for fisheries at the Missouri Cooperative Fish and Wildlife Research Unit of the U.S. Geological Survey. The focus of his research is on predicting how environmental factors influence the structure and function of large river-floodplain communities and ecosystems by defining historical and existing patterns, evaluating the influence of human activities, and applying this to restoration and enhancement of aquatic resources ranging in scale from species to landscapes. He received his

B.S. degree from Cornell University and his M.S. and Ph.D. degrees from Colorado State University.

Gerald E. Galloway, Jr. is research professor at the University of Maryland, College Park. He is also vice president of the Enterprise Engineering Group at the Titan Corporation, Fairfax, Virginia. He also served as secretary of the United States Section of the International Joint Commission in Washington, D.C. Dr. Galloway has served as a consultant on a variety of water resources engineering and management issues to the Executive Office of the President, World Bank, the Organization of American States, the Tennessee Valley Authority, and the U.S. Army Corps of Engineers. Dr. Galloway is a former dean of the Academic Board (chief academic officer) of the United States Military Academy. Dr. Galloway holds master's degrees from Princeton, Penn State, and the U.S. Army Command and General Staff College. Dr. Galloway received his Ph.D. degree in geography from the University of North Carolina.

Robert G. Healy is a professor of resource and environmental policy and public policy studies at Duke University. He also directed the Center for International Studies at Duke from 1994 to 1996. Before going to Duke in 1986, he was a senior associate at the Conservation Foundation–World Wildlife Fund in Washington, D.C. His current research interests are in land use and environmental policy in rapidly growing areas; environmental policy in developing countries; interactions between conservation and economic development; tourism planning and policy; and policy for parks and protected areas. He received his B.A., M.A., and Ph.D. degrees in economics from the University of California, Los Angeles.

Edwin E. Herricks is a professor of environmental biology at the University of Illinois. His research analyzes and interprets the effects of environmental change on species, populations, and communities of organisms in both aquatic and terrestrial environments, with a particular emphasis on the development of methods to improve environmental decision making and ecologically relevant engineering design. Dr. Herrick's specific research areas include biological monitoring procedures; time-related consequences of contaminant exposure; analysis of organism habitat relationships in streams and wetlands; and development of engineering design approaches that minimize environmental impact. He received his B.A. degree in zoology and english from the University of Kansas, his M.S. degree in sanitary and environmental engineering from

Johns Hopkins University, and his Ph.D. degree in biology from Virginia Polytechnic Institute and State University.

Catherine L. Kling is a professor of economics and heads the Resource and Environmental Policy Division of the Center for Agricultural Rural Development (CARD) at Iowa State University. Before joining the faculty of Iowa State University, she taught at the University of California in Davis. Her fields of interest include natural resource and environmental economics, welfare economics, and applied econometrics. In her studies at CARD, Dr. Kling is examining how agricultural practices affect water quality, wildlife, soil carbon content, and greenhouse gases. Dr. Kling received her B.B.A. degree in business and economics from the University of Iowa and her Ph.D. degree in economics from the University of Maryland.

Linda A. Malone is the Marshall–Wythe Foundation Professor of Law at the College of William and Mary, where she has worked since 1988. During her career, she has clerked for Judge Wilbur F. Pell, U.S. Court of Appeals for the Seventh Circuit, and practiced law at Alston, Miller & Gaines in Atlanta and at Ross, Hardies, O'Keefe, Babcock & Parsons in Chicago. She has taught law at the University of Arkansas School of Law and been a visiting professor at the law schools of Duke, Illinois, Arizona, Denver, Virginia, and Washington & Lee. Ms. Malone is the author of several publications, including a treatise *Environmental Regulation of Land Use* and a casebook *Environmental Law*. She is also the associate editor of the *Yearbook of International Environmental Law* and a member of the Advisory Board of the National Enforcement Training Institute of the Environmental Protection Agency. She received her B.A. degree from Vassar, her J.D. degree from Duke University, and her LL.M. from the University of Illinois.

Ram Mohan is vice president and director of coastal engineering for Blasland, Bouck & Lee (BBL) Inc. Dr. Mohan is a registered P.E. in the State of Maryland and has more than 14 years of experience in dredging systems, port and harbor planning, river and channel hydraulics, coastal modeling, contaminated sediments, and environmental planning. He also serves on the board of directors for the Association of Coastal Engineers and the Western Dredging Association, and chairs the Physical Oceanography Committee of the Marine Technology Society. He is adjunct faculty member at Johns Hopkins Whiting School of Engineering and at Texas A&M University's Center for Dredging Studies. He received his

B.S. degree in naval architecture from Cochin University of Science and Technology, India, his M.S. degree in ocean engineering from the University of Rhode Island; and his Ph.D. degree in coastal engineering from Texas A&M University.

Max J. Pfeffer is a professor of development sociology and associate director of the Agricultural Experiment Station at Cornell University. His research has focused on the social aspects of agriculture, the environment, and development planning. He has done recent work on the social dimensions of watershed planning within the New York City watershed, and he works on natural resource management in Central America as well. Dr. Pfeffer was a member of the National Research Council's Watershed Management Committee and the Committee to Review the New York City Watershed Management Strategy. He received his Ph.D. degree in sociology from the University of Wisconsin, Madison in 1986.

Doug Plasencia is a consultant at AMEC Earth and Environmental in Phoenix, Arizona. He also worked as an engineer and vice president of Kimley-Horn & Associates in Phoenix, Arizona. He has more than 17 years of experience in the field of floodplain management and stormwater management working for public agencies and most recently as a consulting engineer, and a professional engineer in Arizona, Nevada, and Virginia. Mr. Plasencia develops watershed- and river-based plans that integrate technology, policy, and implementation into long-term management strategies. He worked with the U.S. Army Corps of Engineers in Sacramento to develop the nation's first nonstructural emergency recovery program for a long-term reassessment of the Sacramento and San Joaquin River basins. Mr. Plasencia was also a hydrologist with the Flood Control District of Maricopa County, Phoenix, Arizona, and chief of flood protection for the Virginia Department of Conservation and Recreation. He received his B.S. degree in forest resource management form the University of Minnesota and his M.S. degree in watershed management from the University of Arizona.

Denise J. Reed has been a professor at in the Department of geology and geophysics at the University of New Orleans, Louisiana, since 1998. Previously she was an associate professor at Louisiana University. Her research interests include sediment dynamics in coastal wetlands with emphasis on sediment mobilization and marsh hydrology, both natural and altered, as factors controlling sediment deposition. Her current research includes sediment dynamics and restoration in Louisiana, the Columbia River estuary, and the Sacramento-San Joaquin delta. Dr. Reed earned her B.A. and Ph.D. degrees in geography from the University of Cambridge, United Kingdom.

Jan A. Veltrop (NAE) is a consulting engineer. He retired in 1994 after 37 years with Harza Engineering Company where he served as chief engineer, executive vice president and director. Mr. Veltrop worked on hydroelectric dam projects around the world, including India, Pakistan, Taiwan, Iran, China, Israel, and the United States. During a leave of absence from Harza, he served as Dean of the Faculty of Engineering at the University of Nigeria, Nsukka from 1964 to 1967. Dr. Veltrop was chairman of the U.S. Committee on Large Dams (now USSD) and President of the International Commission on Large Dams. Dr. Veltrop received the Rickey Medal from the ASCE in 1997 for his contributions in the field of hydroelectric engineering. He also served as a Commissioner of the World Commission on Dams in August 1998. He received his B.S. degree from the Technological University of Delft, Netherlands; his M.S. degree from Rensselaer Polytechnic Institute; and his Ph.D. degree from the Massachusetts Institute of Technology.

STAFF

Jeffrey W. Jacobs is a senior program officer at the Water Science and Technology Board of the National Research Council. His research interests include organizational and policy arrangements for water resources planning, water resources science and policy relations, and river system management. He has studied these issues extensively in Southeast Asia's Mekong River basin and in the United States, and has conducted comparative research between water management issues in the United States and Southeast Asia. He received his B.S. degree from Texas A&M University, his M.A. degree from the University of California (Riverside) and his Ph.D. degree from the University of Colorado.

Ellen A. De Guzman is a research associate with the Water Science and Technology Board of the National Research Council. She has worked on a number of studies including Privatization of Water Services in the United States, Review of the USGS National Water Quality Assessment Program, and Drinking Water Contaminants (Phase II). She co-edits the WSTB newsletter, annual report, and manages the WSTB homepage. She received her B.A. degree from the University of the Philippines.